HOW TO BUILD A FLYING SAUCER

A Beginner's Guide to Gravity Amplification Systems

PREFACE

Hello reader, this book previously existed as a chapter in my last book, Alien Revelations.

Alien Revelations was my attempt to share my experiences as an alien abductee along with everything that I learned from those experiences.
When I say things that I learned, I mean that I received very specific information concerning the fundamental mechanisms that govern our reality, from the nature of existence before the birth of our Universe, to revelations about the nature of Gravity, and the strangeness of the Quantum world.
I also go deep into the nature of consciousness and even talk about what happens to us when we die.

This book will be much shorter as it is focused on one subject. In the following pages, I will share everything that I learned about the Extraterrestrial's method of manipulating gravitational fields during my time with them.
If you wish to learn more about my encounters, I would recommend that you read my first book, but if you simply want to know the secret of gravity manipulation, then this book will contain everything that you need.

At the time that I wrote my first book, I believed that the information held in these following pages was potentially too destructive to be released to the public.
The potential impact that this information could have on the

human race is colossal. It has the power to liberate us by allowing us to expand out into our Universe both cheaply and efficiently.

Unfortunately this technology like all others has potential for abuse. If this technology was to be wielded as a weapon, it could spell the end of our entire species.
The reason that I have decided to finally release this information is that I strongly believe that this technology is about to be released to the public from a very unlikely source.

In the twenty years since I was first exposed to this information, I would never have guessed that I would be inspired to release it to the public by the actions of a rock star.

I am referring to the actions of former "Blink 182" vocalist, Tom Delonge. In an extremely short amount of time Tom has had a huge impact on the Ufology community. He seems to have used his media profile as a kind of crowbar to prise apart the seams of the wall of silence that has been erected by secretive organisations within governments around the world.
He has even managed to get the US navy to admit that our skies are being trespassed by unknown, advanced, technological aircraft.

His latest work seems to be focused on unlocking the secrets of a small piece of metal alloy that he claims came from an ExtraTerrestrial spacecraft.
I'll jump straight to the point and say, I believe that he is right and that he is very close to making a discovery that will change the course of human history, forever.

Even if he is unsuccessful, he has already left enough clues for someone else to pick up the baton and cross the finish line by making a monumental discovery for themselves in a relatively short amount of time.

I hope that Tom is successful and I applaud his achievements, but I am worried that he has made some powerful enemies along the way. He has made a lot of progress in a short period of time, which I believe may have made him a target of some very unscrupulous people.

I strongly desire that this technology be released to the public, but until now I have been too cowardly to take the responsibility of releasing it by myself.
Now that it seems likely that Pandora's box is about to be opened, I can take the opportunity to democratise this new technology.

If the release of this new technology is anything other than completely open source, I believe that it will lead to disaster for the human race.

If we try to continue to keep the lid on this technology, even partially; then I believe that the imbalance of power could lead to a global catastrophe.
My concern is that any organisation that invests time and money into solving the riddle of Extraterrestrial propulsion, will wish to profit from any discoveries that they make and it is this course of action that could spell disaster for all people.

History has taught us that new technologies that have military applications often lead to more bloodshed until those technologies become commonplace.

Would allied forces have dropped hydrogen bombs on Hiroshima and Nagasaki, if Japan also had access to that same technology.

As ironic as the acronym for Mutually, Assured, Destruction is, it has prevented or global superpowers from utilising these weapons again since that fateful day in 1945.

My hope is that this book will not only democratise gravity manipulation technology, but it will also speed things along a little. Hopefully this may be enough to prevent hidden organisations from blocking its release to the public.

What will follow, is a complete breakdown of my own understanding of the propulsion systems that the Extraterrestrial visitors use to traverse vast distances.

From my own point of view this book represents putting all of my cards on the table of an incredibly high stakes poker game

One of the reasons for my change of heart concerning the release of this information may also be due to my increasing pessimism as I get older. It feels as if our global societies are becoming more and more polarised politically and I am under the constant impression that we are on the brink of another massive global war between our world's political superpowers. The reason for not releasing my insights concerning gravity amplification is that it could give small, more unstable governments access to a cheap and easy to build superweapon that would be the rival of our most powerful weapons of mass destruction. The problem is that as time moves on and our more stable superpowers move closer to the brink of war, this concern becomes less and less relevant.

I now find myself in a place where the pros of releasing this information seem to outweigh the cons.
The immediate benefits of this technology are obvious. The ability to cheaply and efficiently explore and expand into space would have a colossal effect on our species. Less obvious benefits include the sudden ability to efficiently fuse hydrogen for power generation and it could also allow us to communicate clearly over vast distances by utilising this technology's ability to manipulate space and time.

If we are able to work together as a species, this new technology could give us the opportunity to enter technological renaissance period and kickstart a new golden age for our species. If we cannot resolve our differences, it could be the catalyst that triggers our own extinction.

That is all pretty heavy stuff and you can see why I was so reluctant to share this information. The problem is that simply knowing this information is a huge responsibility. I am either the person that denies our species the opportunity to thrive and expand by concealing it, or I am the guy that put us on the path to complete and utter annihilation by sharing it.

I have come to the conclusion that there must be a reason that I was given this information and that reason can only be to either use it or share it.

Since I lack the technical proficiency to use it by building an actual gravity amplifier, my only other option is to share it.

I had considered actually trying to build a gravity amplifier, so that I could perhaps exercise some level of control on how this technology would be used, but that seems extremely optimistic. I doubt that I have the ability to learn the necessary skills to build one of these devices. Even if I did manage to construct such a device, how exactly could I direct how it would be used?

After waiting for more than 20 years I have come to the conclusion that perhaps it is time to simply roll the dice and release this information to see what happens.

I will start by simply explaining the method to manipulate gravitational fields in the first chapter and then I will go on to provide a history of work that has been done in this field that reinforces the likelihood that this method is correct. After that I will highlight UFO sightings and encounters that best demonstrate evidence that this is the same method of gravity manipulation that our Extraterrestrial visitors use. I will then end with a short Afterword. This will be a very short book, but I believe that this is the best way to get this information out to the public.

I will also be providing diagrams to help illustrate the mechanisms of this technology.

CHAPTER 1:

Gravitational Manipulation Method

This method of gravity manipulation takes advantage of the fact that when matter reaches a velocity within a certain percentage of the speed of light, its mass increases and the way that it interacts with Space and Time alters.

This idea was first introduced by Albert Einstein when he published his Theory of General Relativity in his 1905 paper "The Electrodynamics of Moving Objects", but thanks to modern advances in computer and satellite technologies it has become a necessary consideration when designing and constructing both of these technologies.

This is all just to say that the main conceit in the method that I am about to share is firmly grounded in known Science.

Thanks to Einstein's work, we know that when matter moves at a

high enough velocity, that it has the effect of reshaping the geometry of space and time around itself. This seems like a good place to start if our intention is to reshape Space and Time with the goal of manipulating gravitational fields.
A big hurdle is that in order to take advantage of this effect, we first have to accelerate matter to an incredibly high velocity before we are able to observe any measurable effect on the geometry of Space and Time.
A loophole can be found if we consider alternative forms of motion. The most obvious one is to move along a single path in a straight line, but this form of kinetic force is the least useful when searching for a solution to our problem of generating <u>static</u> gravitational fields.

Another possibility is moving in circular paths of motion, and you would be forgiven for thinking that is where I am leading you considering the importance of circular paths of motion in my previous book.

The specific form of motion that I am thinking of is vibration.

When we talk about vibration or oscillation we rarely measure this form of motion the same way that we measure linear motion; in meters per second or miles per hour.
We tend to measure <u>frequency</u> of vibrations and the <u>amplitude</u> of those vibrations.

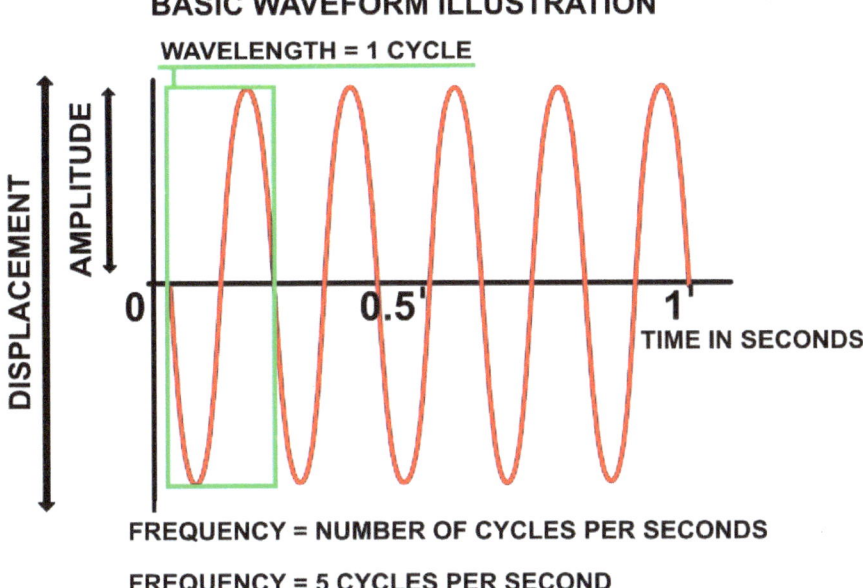

Frequency simply measures the number of motions per second. As we consider smaller and smaller quanta of matter, the measurement of meters per second become less and less useful.
We instead consider the amplitude of the observed vibrations which gives us a measurement of the amount of motion between oscillations. The linear velocity of the vibrations is never really considered, but for our purposes, it is an extremely important value. For example, if we were to strike a tuning fork, how fast is the tip of that fork moving through space? Instead of concerning ourselves with that value, we measure the frequency of the motions in Hertz and the distance that they move as Amps. We could measure the velocity of the fork tip by multiplying the frequency of the waves with their wavelength, which sounds dangerously like math, so let's quietly move on shall we.

The tips must be moving pretty fast because they move back and forth at several hundred times a second which has enough force to strike air molecules hard enough to generate a prolonged sound. This is a simple example of the kinetic forces involved when matter oscillates, but there are much more efficient ways to make matter vibrate at much higher frequencies than striking a tuning fork.

It is important to remember that all matter is constantly oscillating naturally thanks to the fact that it is made up of oscillating units of force.

Whenever we add energy to matter in the form of heat, kinetic force and electrical energy we can observe that the molecules that make up that matter oscillate at higher frequencies and increased amplitudes. This effect has countless applications in modern technology whenever we convert electrical energy into heat sound or light or vice versa.

For our purposes, we simply want to increase the amount of kinetic energy within a single unit of matter. If we are able to make that matter oscillate fast enough, perhaps we can generate enough velocity to generate a measurable effect on the Space and Time around it.

The most obvious method of generating increased kinetic energy from vibration within a volume of matter is by applying large amounts of electrical energy to that matter.

The application of electrical charge to matter is not quite as simple as it sounds. Electricity only flows when there is a difference in charge between two points. Electricity likes to flow through matter and getting it to simply occupy a volume of matter involves some trickery. Luckily for us there is a common method for doing just that. This common method utilises a device called a capacitor.

A capacitor is an electrical component, usually constructed from two plates of conductive matter that are separated by a non conductive plate called a dielectric.

Capacitor Diagram

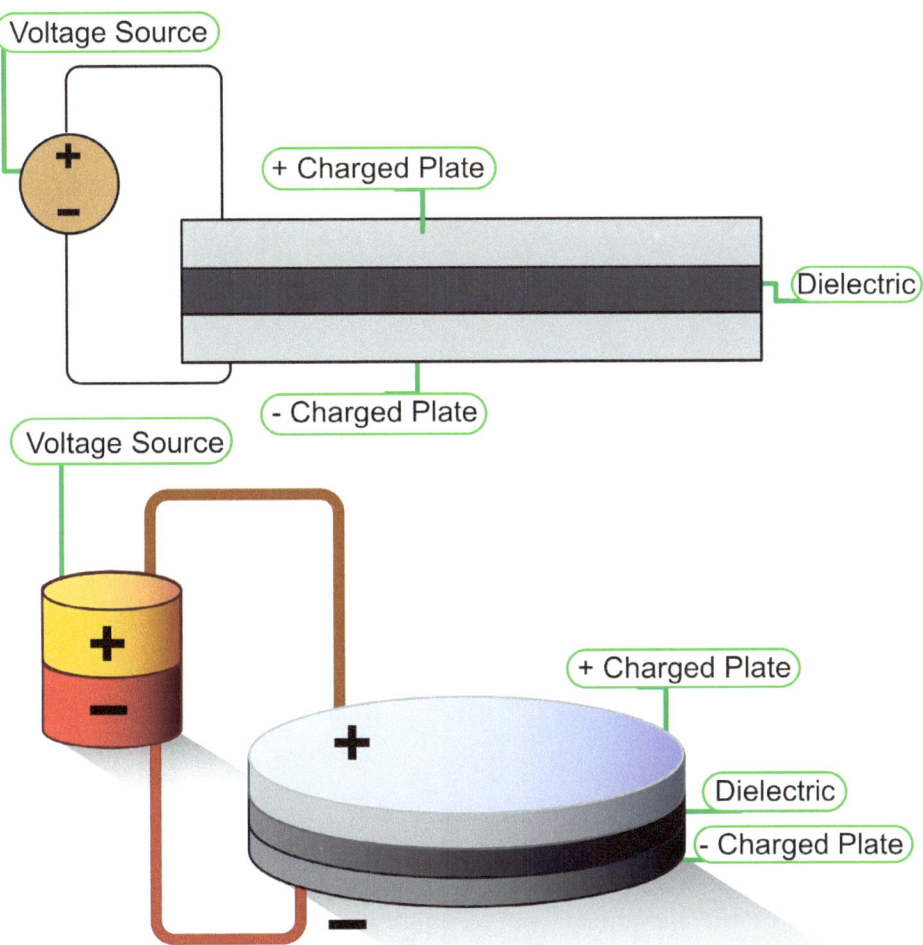

When a capacitor is placed within an electronic circuit, the positively charged plate fills with positive, electrical, potential energy while the opposite plate fills with a negative charge. The result is that the atoms and molecules within those plates become excited and oscillate at high frequency. This can often lead to a high pitched whine as the plates accumulate charge. If you are old enough to remember old flashbulb cameras, you will be familiar with this sound. The sound is generated by the motion of the molecules within each plate of the charging capacitor within the camera, as they bounce around at high frequency due to the presence of the electrical charge.

I believe that the kinetic energy generated by oscillating molecules within a capacitor has the potential to create miniscule gravity waves.

Now if I am proposing that this increase in kinetic energy within a capacitor is sufficient to generate a measurable effect on Space and Time then you may be asking yourself, why have we never detected such an effect before?

I believe that we are exposed to minute, high frequency gravity waves often, but we lack the ability to detect them.

The problem with gravity waves is that they are waves that in a real sense do not exist. They do not have a mass and unlike every other kind of wave they propagate through a medium that has no physical form.
A gravity wave moves through the fabric of space, but unfortunately space is not a fabric it is closer to a concept. It is the background value against which we measure all physical phenomena. In my previous book I explained how and why this value is the speed of light, because photons represent our perception of the seemingly unchanging phenomena that is the Protoverse.
Luckily for us that information isn't really important for this method, but if you are curious, I give a full explanation of the Protoverse in my previous book "Alien Revelations." When we observe phenomenon that comes close to violating this background value (the speed of light) by matching it, it is then reshaped by

its desynchronization of motion with all other macro scale phenomena.

The desynchronized matter no longer exists within the shared medium of Space and Time and begins to experience changes to the way in which it interacts with reality.

When matter moves close to the speed of light, it is moving away from the shared medium of Space and Time because in a real sense it is effectively moving toward a single state existence, like the Protoverse.
When I say single state I mean that externally the Protoverse seems to be locked into a repeating cycle of motion, but internally it is infinitely dynamic and complex.

When matter moves close to the speed of light, this desynchronization has the effect of reshaping the flow of dark energy (Expansion) around the fast moving object, similar to a gravitational field. Mass reshapes dark energy, because it works against Expansion, because matter is essentially condensed dark energy. When matter moves through Space close to the speed of light, it works against Expansion along a single vector, therefore the effect is similar to that of a powerful gravitational field.

Now if an object could be made to move close to the speed of light, without actually changing its physical position, then it would be able to generate a gravitational effect along a single axis.

The issue is that the effect would be so minute that it would be all but undetectable. When a single atom moves back and forth at an extremely high frequency, it will generate a small gravitational wave similar to how a vibrating object would create waves in a small pool of water. The difference is that water has mass and is visible, whereas space possesses neither of these attributes. Space is merely the background against which we measure all phenomena, so what is actually changing is the local physical dimensions in an area of space around the vibrating atom. The waves propagate outwards with the flow of Expansion, which is also why observed Gravity Waves seem to move at the speed of light.

The scale and the force of the small Gravity Waves are incredibly minute and as soon as the waves strike anything with mass, the effect is defused almost immediately. This means that any device that we construct to measure these waves would have the undesirable effect of eliminating the phenomenon that it was designed to observe.

Apparatus does exist that has been designed to observe Gravity Waves, but not at this scale. Interferometers are huge Gravity Wave observatories that monitor changes in long range beams of laser light to record any changes to the curvature of Space/Time. Another kind of Observatory called a Resonance Mass Antenna, monitors changes in a large isolated mass's resonant frequency, to observe the effects of gravity waves as they pass through matter.

Both of these solutions are geared toward the detection of large scale low frequency Gravity Waves emitted by large scale events such as colliding Black Holes over vast distances, neither of which help us to detect small scale gravity waves.

What we are left with is a hypothetical wave that is so small and weak that it is pretty much undetectable and yet I am proposing that these waves are the driving force of the incredibly fast and manoeuvrable craft that we see extraterrestrials piloting.

So now I will tell you how to amplify these waves and why the humble capacitor may become the most important technological device imaginable.

The weakness of small gravity waves represent an engineering problem that the Visitors have solved and I will now tell you how.

The main issue with the small Gravity Waves, is that they are almost instantly cancelled out either by the mass of other matter that they strike or similar waves that may be moving in the opposite direction. This is a familiar problem for anyone that works within a field that deals with wave interference, this could be electromagnetic waves or even oceanic, the concept remains the same.

The solution can be found by finding a way to combine the waves so that they work together to create a single wave force. In the field of acoustics, the best way to do this is to find the harmonic resonance of the object that is being used to create the desired sound.

The harmonic resonance is a special frequency of vibration that travels along the surface of a material in such a way so that each wave works together to reinforce the structure of its neighbours. The structure of each wave is perfectly symmetrical, so that the forces working on it maintain its prolonged existence. Different materials have different harmonic resonances, so to take advantage of this effect, it is necessary to find this magical frequency for each one.

We are not working with sound waves but gravity waves which require incredibly high frequency waves, but luckily for us, harmonic waves can be multiplied many times to generate much higher frequencies. These miraculous Frequencies have mathematical properties that allow them to be multiplied several times, to produce much higher Frequencies.
When harmonic waves of sound are produced, the waves combine to generate a much more powerful wave force and the same is possible for Gravity Waves.
If we were to construct a capacitor from a conductive material with a particularly high harmonic resonance, we would be able to increase that resonance by multiplying it many times. If were able to increase it enough, it could be possible to create a gravity

wave force that combines rather than cancels itself out.

At this high frequency, it would be possible to tune the amplitude of the oscillations, so that the atoms within that material are moving as fast as possible for that frequency. If the amplitude is high at a high enough frequency, then the atoms would be moving a greater distance in a short amount of time. In other words they would be moving faster, hopefully fast enough to affect Space-Time.

So we are trying to oscillate the atoms within our material, at the highest possible Frequency and at the highest Amplitude possible in order to create Gravity Wave sources throughout that material.

When we consider that this wave force is being created by the individual atoms within the material of our capacitor, we may realise that these miniscule waves are being produced by trillions of single points all working together all at once. In this way we have turned an undetectably weak wave into a powerful linear wave force.
Now that we have done this, we can focus those waves upon a single point in space to produce a linear Gravitational field.
The way that we do this is simply by designing our capacitor to be a parabolic dish shape.

A parabola is a simple symmetrical curve and the benefits of using this shape are that every point on the inner surface of our bowl is at 90 degrees to a single point in space. This means that any waves emitted from the inner surface of our capacitor intersects at a single point in space. This focusing of our waves produces linear pressure on an area of space, which in turn creates a linear gravitational field, a phenomenon that does not exist in nature as far as I know.

Gravity Amplifier Illustration

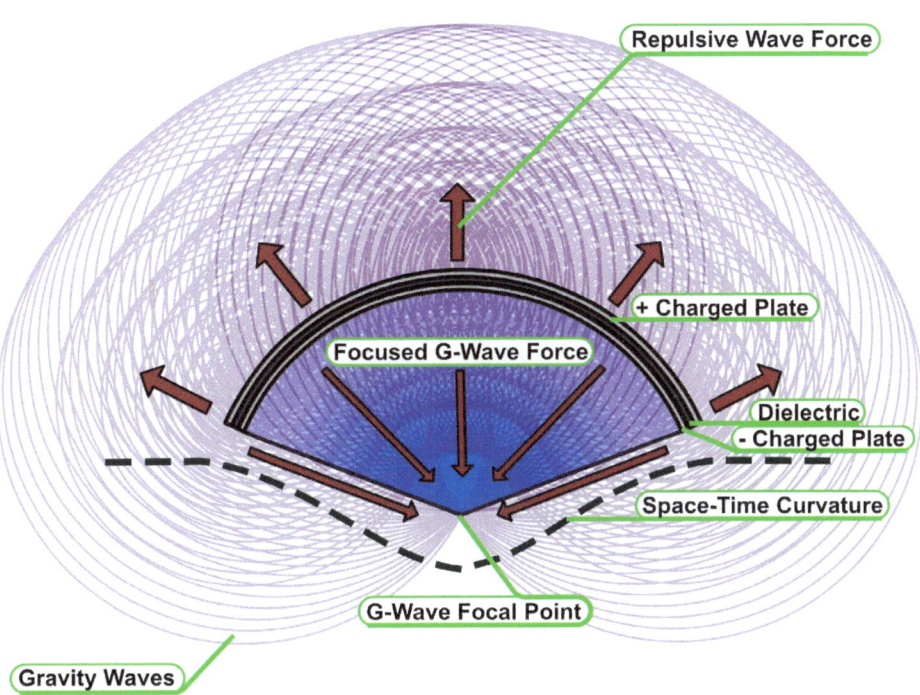

If everything that I have written here is correct then what we have here is an actual method for manipulating gravitational

fields and it is therefore is one of the most important pieces of text ever written in History. Or it would be if this was the first time that this had ever been done.
Unfortunately I do not believe that it is and this text is really just the first time this information has been released to the public.

I believe that this technology already exists and that working gravity amplifiers have been built as far back as the 1940s.

I will return to this subject soon, but for now let's concentrate on the specific mechanics of our gravity amplifier.

Earlier I alluded to the fact that small Gravity Waves have the unfortunate tendency to be cancelled out by other similar waves and also when they strike against anything with mass. This makes things tricky if we are trying to produce a coherent combined wave force with which to work with.

Once again the solution can be found in the specific traits of the humble capacitor.

In order to make use of any gravity waves that we manage to produce, we have to make them all flow in the same direction, at the same time.

We can already control their timing by taking advantage of the harmonic resonance of the material that we are using for our capacitor plates. Fortunately the direction or axis of the vibration of the atoms within the surface of our capacitor can also be controlled thanks to the way that capacitors work.

Now I may have misled you a bit at this point by making you think that the best way to oscillate the molecules in a material is by exposing it to large electrical charges, that is technically correct, but the best way to control those oscillations is by placing our gravity wave source adjacent to a powerful electrical charge.

When working with simple capacitors we use two adjacent conductors and separate them with a non-conductive material, called a dielectric. The first conductive plate in our circuit is flooded with negatively charged ions, whereas the second plate becomes dominated by positively charged ions. This second plate will become our gravity wave source.

The first plate contains the most energy in the form of electrical charge, but this charge also has a dampening effect on the amplitude of our vibrations.

The saturation of electrons within the first plate prevents the atoms within the material of that plate from reaching a sufficient amplitude to generate Gravity Waves.

Imagine a dance floor filled with super energetic dancers, all whirling around and moving back and forth at high rates of speed. As the dance floor fills up, the motions of the dancers becomes limited because there is not enough room to perform the more ambitious moves.

Demonstration of Increased Activity With Fewer Actors

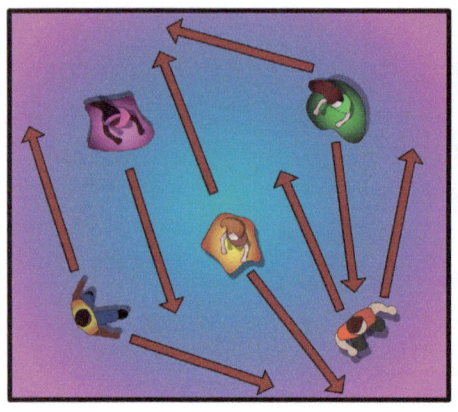 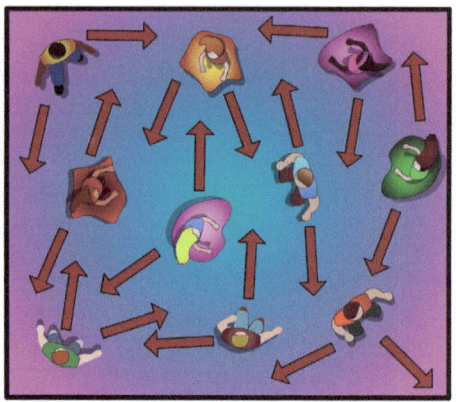

This is similar to how the saturation of ions within the first plate limits the motion of the atoms within it. Every extra electron repels the other electrons and negatively charged atoms around itself and although this repulsion does add energy to the material, it also limits its free motion.

Within the second plate however, the atoms are motivated by their lack of electrons and so they have much more room to move around. Another bonus is that atoms have much more mass than electrons, so their motion has a much greater effect on Space-Time.

We can even affect the axis that the atoms follow in their oscillations as they will move toward and away from the first plate. This is because the motions of the molecules within the second plate are purely guided by their attraction to the powerful opposite charge on the first plate. Whenever a charge is suspended by a capacitor, the powerful static charge creates a magnetic field and it is the lines of force of this field that dictates the motions of the molecules within the positively charged plate.

The oscillations of the first plate are created by the flood of excess electrons, which cause the atoms within it to bounce around at random. Within the second plate however, the atoms within it become agitated by the lack of Electrons and their motions are generated by the magnetic attraction of their composite Protons. The more powerful the charge, the stronger the magnetic field and with careful tuning of the charge and materials used it is possible to generate enough controlled vibration within the second plate to produce Gravity Waves.

In order to focus the Gravity Waves at a single point in space it is necessary to build our capacitor following a very specific blueprint. As mentioned earlier, the positively charged plate should be shaped into a parabolic dish shape. This is so that as many Gravity Waves as possible will be focused on a single point in space.

The negatively charged plate should then be shaped so that it provides a target for the oscillations of the positively charged ions in the positive plate. We are presented with two options here. We could either craft each plate into an extremely thin layer and stack them like pancakes, separating each layer with a dielectric (insulator). This method would be very difficult as it would require an extremely highly skilled level of engineering, both to craft each layer and also to bind them together.

The reason that each layer would have to be so thin, is because we want to eliminate any motion within the positively charged plate that is not moving along our desired axis, as this motion would produce Gravity Waves that would conflict and cancel out the focused and harmonized waves that will produce our Gravitational force.

Diagram of Thinly Stacked Capacitor Plates

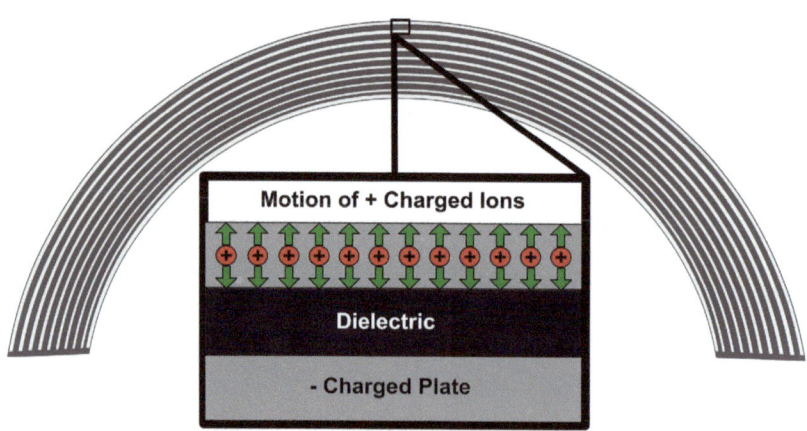

The second and much easier method is to craft the negatively charged plate into something like a sphere and then place it at the focal point of the inner surface of our positively charged plate.

The reason that we need to specifically use a sphere, is so that it is either of equal mass or more to our dish shaped plate while also placing as much of its mass within the focal point of its opposite plate as possible.

Bell Configuration Gravity Amplifier

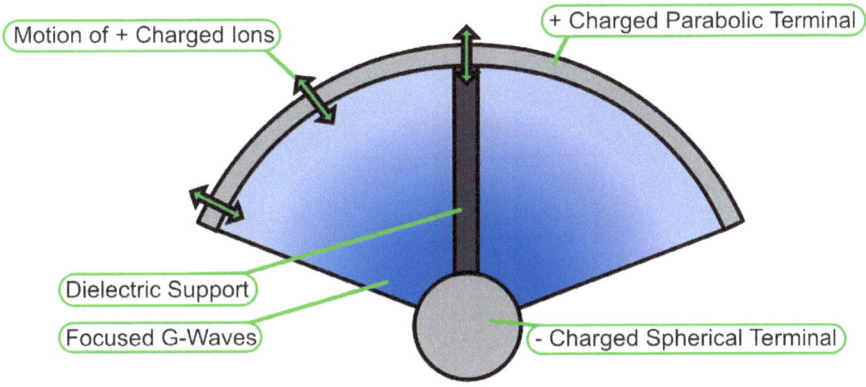

This method is much less efficient, but is a much easier way to produce the same, if somewhat weaker effect. The relatively large distance between the plates does improve the accuracy of the axis of atomic oscillation, but the magnetic force between them is also much weaker, causing the oscillations to be weaker and less accurate.

From now on I will refer to both of these methods as the G.A.L method of gravity manipulation, as G.A.L is an acronym of Gravity, Amplying Lense. This is accurate because the surface of the device produces and amplifies gravity waves and its shape focuses them into a linear, coherent force, as a lense would.

So now we I have explained the mechanisms of the Gravity Amplifier, but I have provided no evidence that such a device is possible. I have given an explanation for something that so far does not exist. Until this device is constructed and these concepts are proven, this whole chapter is no more than a very technical and abstract fairy tail.

I will now provide evidence that not only is it possible to construct this device, but also that these devices have been constructed in laboratories on Earth since the 1940s. I will also provide some evidence that this is also the basis for the technology that the visitors use to travel vast distances between stars.

CHAPTER 2:

Gravity Amplifier, Historical Evidence

My own introduction to the subject of Gravity Amplification began in 1996. The incident that started my journey occurred whilst I was watching a speculative Science TV show called "Future Fantastic". Every episode featured three examples of research in new and exciting technologies that could potentially have a huge impact on our society. Each episode centered around a different theme, for example one episode highlighted advancements in Artificial Intelligence and robotics while another centered around biological and genetic enhancements. The episode that put me on the path to writing this book however was based on the theme of advanced propulsion and aeronautical systems. This episode covered things like ionic propulsion systems and nuclear powered rocket engines, but the segment that really got my attention featured the work of a man named Thomas Townsend Brown.

While I watched the show, I saw grainy footage of Mr Brown's experimental apparatus. As soon as I saw it, I had a very strong feeling of recognition. I knew that I had seen this device before and I also immediately understood some elements of the mechanisms that made the thing work.

Over the years since, I have been able to piece together a more complete picture of where I gained this secret knowledge and the specifics of how the apparatus that I saw in that old grainy footage functioned.

For information on how I gained this knowledge, you will have to read my last book "Alien Revelations," but for the specifics of how this device works, I will share that next.

Thomas Townsend Brown spend a large portion of his life from the early 20s until his death in 1985, trying to find a link between Magnetism and Gravity. He named this field of research "Electrogravitics".

In 1921 Thomas believed that he had discovered a link between electromagnetic fields and gravitational fields while using a piece of apparatus called a Coolidge tube.

As a young man, Thomas would spend a lot of his time conducting experiments in his own private laboratory that his parents had built for him to supplement his education in his favorite field, Physics.

One of the pieces of equipment that Thomas had access to was a piece of medical equipment called a Coolidge tube.

A Coolidge tube is an essential piece of medical equipment that is used to emit X-rays for X-ray photography. It is constructed from a bulb of glass, that has had all of the air removed to create a small vacuum. Within the bulb are two components, an anode and a cathode. These are similar to the components of a capacitor in that the cathode holds a negative charge, while the anode is filled with a positive charge.

These components are asymmetrical as they have different roles to the electrode plates of a capacitor. In the Coolidge tube, the cathode's purpose is to emit electrons, that will strike the anode.

This interaction then causes the anode to emit high energy electromagnetic waves, specifically, X-rays.

To fulfil this purpose the cathode is made from a coiled, negatively charged, thin wire and the anode is made from a thick disc of tungsten that is positively charged. The thin wire has a lot of surface area in order to release electrons that can then hopefully strike the oppositely charged anode disc. The anode is a larger disc, partially to provide a larger target for the emitted electrons, but also the extra mass ensures that most of the elections do not simply pass through the anode without striking an atom within it, as it is the collision between the electrons and the tungsten atoms that provide the high energy interaction that generates the X-rays.

Coolidge Tube Diagram

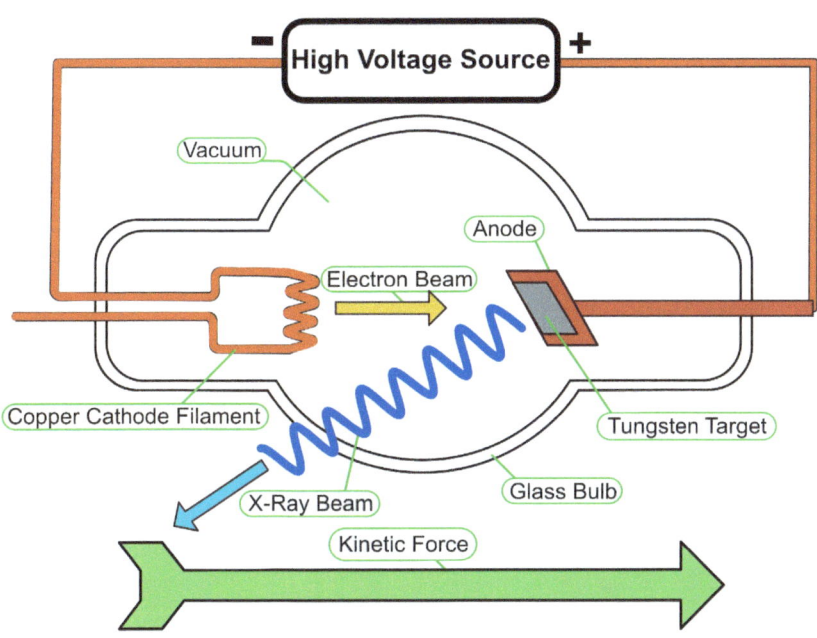

If we look at the Coolidge tube with the knowledge that the arrangement of a large positive charged terminal that is separ-

27

ated across a small distance from a smaller negatively charged terminal is close to the desired set up for a simple Gravity Amplifier as illustrated in the previous chapter, then we might be able to guess why Thomas Townsend Brown became famous for his efforts to find a link between magnetic fields and gravitational fields.

To us, it is not surprising that when he connected his Coolidge Tube to a high voltage source, that he noticed that it would inexplicably move in the direction of the positively charged anode.

This discovery then kickstarted a lifelong quest for Mr Brown that would unfortunately end unsatisfactorily with his death of old age without ever being resolved.
Thomas Townsend Brown worked as a scientist in various capacities, but he would keep returning to the mystery of the moving Coolidge tube throughout his life.
Over the years he built several different versions of his Electrogravitic apparatus. The earliest version was a crude capacitor, that was suspended from a swinging pendulum. When connected to a high voltage source, the capacitor would move around a central pivot seemingly proving a connection between electromagnetic fields and gravity. Unfortunately this apparatus would also produce a different motivating force known as ionic wind, which undermined Brown's own electrogravity hypotheses.

Brown would even go as far as to patent a redesign of this device to make better use of the ionic wind effect, adding ionising vanes to the leading edge of the capacitor to ionise the air in front of it, but he never stopped trying to find a link between gravity and electromagnetism.

Electrokinetic Apparatus, Utilizing Ionic Propulsion

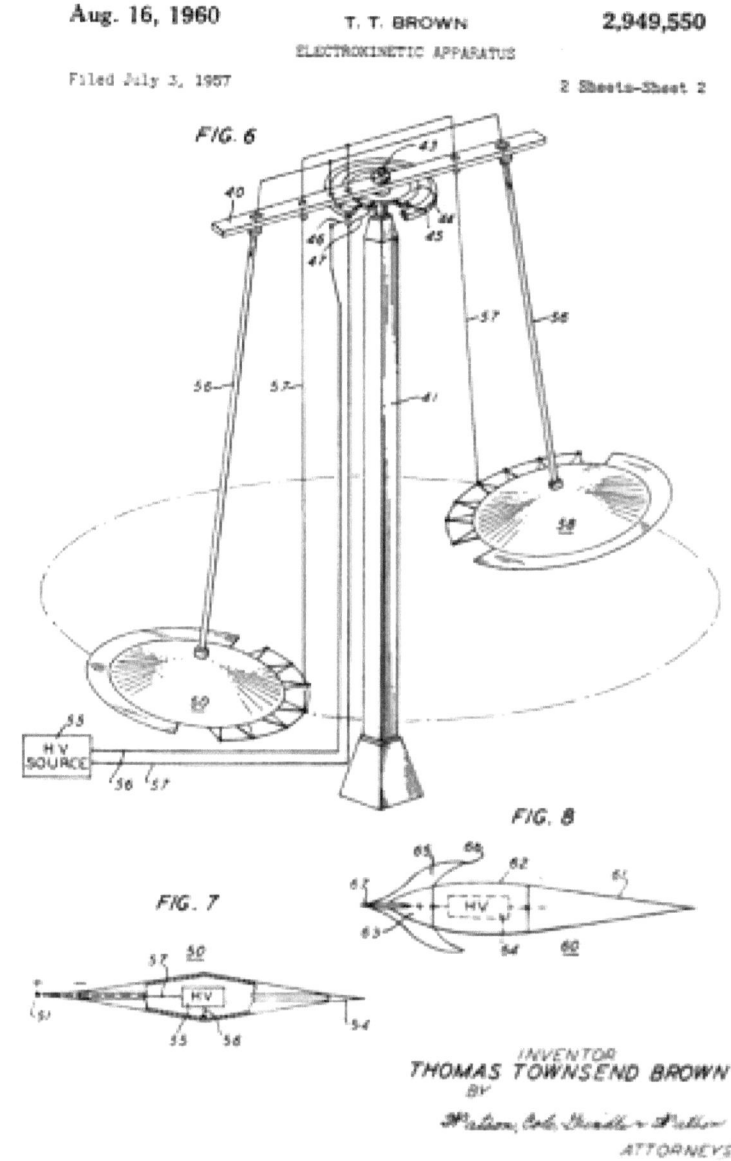

The TV show "Future Fantastic" showed examples of Brown's later experiments, at a stage of his life when his apparatus had undergone many redesigns. Over the years Brown had found that his apparatus would generate more kinetic force when the positive terminal was much larger than the negatively charged terminal. He had also found that the kinetic effect was magnified when the positive terminal was shaped into a parabolic disc and the negatively charged terminal was placed at the focal point of the positively charged disc.

The TV show played footage of these parabolic disc shaped capacitors bouncing around in Brown's laboratory.

This meant that what I saw on the TV screen in 1996 was a perfect demonstration of a mechanical principle that I believe had been explained to me at some earlier point in my life.

As I watched the screen I could feel an unmistakable sense of familiarity. It was like déjà-vu, only much deeper. I instantly recognised the device that I saw in the old footage of Brown's disc shaped apparatus. In my mind's eye, I could see a representation of the parabolic bowl suspended in a 3D representation of the way that the device was reshaping the space around it.

Spatial Distortion Generated by G.A.L Unit

HOW TO BUILD A FLYING SAUCER

The image appeared in my mind so briefly that it could only have come from my subconscious mind. I was familiar with these kind of flashes of recall as I had experienced them throughout my life. They would usually contain images of the visitors in various situations. The memories rarely made sense, but this one seemed to revolve around something even more abstract. I was instantly aware that this memory could contain important scientific information and I became obsessed with uncovering that information.

In that initial moment of recognition, I was aware that Thomas Townsend Brown's device was focusing some kind of force into a single point that was exerting some kind of pressure on Space itself.

Immediately after watching the footage, I tried to learn as much about electromagnetism and gravity as I could. I seemed to have been given a glimpse of a secret that I intended to uncover.

Unfortunately everything that I then learned seemed to tell me that what I had seen was impossible.

The first problem is that space has no physical form, therefore finding some way to physically exert pressure on it seemed impossible. The second problem was that gravity waves only seem to be produced by astronomical events such as colliding neutron stars. My biggest problem was that if gravity waves could be produced simply by exposing matter to high voltages, then surely we would have discovered them at some point in the nearly two centuries since Micheal Faraday first learned how to induce electrical current in 1821.

In the years since first seeing Brown's experiments on TV, I put aside my own theories and resolved to simply observe any more instances of the specific déjà-vu that I had felt while watching "Future Fantastic".

I began to almost exclusively watch shows on the Discovery channel. I would also take out books on cosmology and physics from my local library. Almost overnight I became a huge science nerd. My interest in hard Science was somewhat undermined by my new UFO obsession that had been kicked off in the previous year when I first realised that all of my bizarre childhood, nocturnal memories all fell under the umbrella of close encounter experiences.

My new obsession led me on a journey that would eventually lead me to write both this book and my previous one. Over the years I began to construct a new hypothesis concerning the nature of reality, but that journey began with revelations concerning the mechanisms of gravity amplification.

I would look for anything that tickled the sense of déjà-vu that I felt when I saw Thomas Townsend Brown's archival footage. Each incident of déjà-vu was another piece to add to the puzzle. Over time a more complete picture started to be resolved and I became obsessed with Einstein's theory of relativity. I knew that it held the key that would unlock something in my own subconscious memory that was just out of my reach. When I heard the story of how he almost scrapped his theory because it depended on the Universe being in a constant state of expansion or contraction, I felt the same flash of déjà-vu that I had felt back in 1996. From that initial disorienting moment of near recall, I was able to build a new perspective of reality that was reliant on Einstein's original work. As I worked on my new model of reality, I suddenly realised how the theory of relativity held the key to explaining how Brown's electrokinetic apparatus had the power to exert pressure on Space and Time. I instinctively knew that the vibration of the molecules within the apparatus were the catalyst for the Gravity Waves, but I didn't know why. I wish I could explain why I knew, it was just something that I felt rather than knew logically. This gives you an insight into how my déjà-vu works. I will learn some obscure scientific concept either by reading about it or watching Science shows on TV and I will simply know things that are at the periphery of the concept that were not explained when I learned about it. The oscillations of the molecules within Brown's floating discs was one of these intuitive insights. Once I applied Einstein's theory of relativity to this insight, I realised that the velocity of the vibrating molecules could have an effect on the curvature of Space. As I built my new theory of everything, I also began to find evidence that my initial flash of recognition concerning Thomas Townsend Brown's experiments were based on something real.

Moving on from the work of Thomas Townsend Brown, I believe that there are more examples of work that has been conducted using the technique outlined in the previous chapter to manipulate gravitational fields.

Another, somewhat more mysterious example in history, where it would seem that this technology may have been worked on, could have occurred at a secret Nazi weapons research site during the second world war.

I am referring to a device known only as "Die Glocke" or "The Bell" in English.

Not a lot is known for certain about this device. There are several anecdotal accounts about it, that seem to mix mysticism and alchemical pseudoscience, but if we strip away the story to its bare bones, there may be something of interest in it just from the fact that this was supposedly an antigravity device called "The Bell".

"The Bell" was supposedly developed at an advanced weapons research base located in a mine at Ludwigsdorf in Germany.

When I first heard about this device many years ago, the only thing that caught my attention was its name and its supposed purpose.

I believe that the most achievable method of gravity manipulation is to oscillate material at incredibly high frequencies, utilising that material's harmonic resonance to achieve higher and higher frequencies and amplitudes.

I also believe that one of the simplest configurations for such a device is to craft a large, positively charged parabolic dome like electrode, with a smaller negatively charged, spherical electrode, suspended at the point of focus of the larger parabolic electrode.

G.A.L Bell Configuration

HOW TO BUILD A FLYING SAUCER

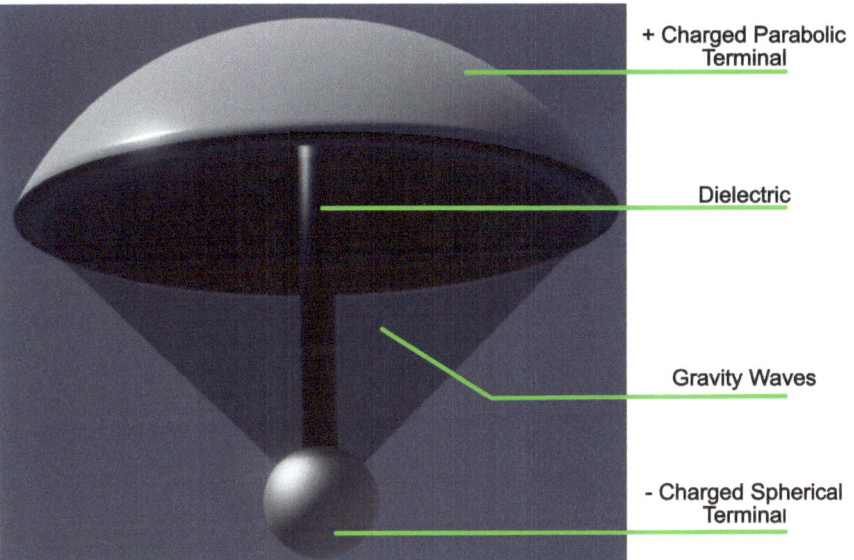

If we consider that name "The Bell," we might immediately notice that the simple G.A.L configuration, does look remarkably like a bell, with a bowl shaped body and a large spherical clapper suspended below. When we also consider that this device utilizes harmonic resonance to generate a perceivable effect on Space-Time, we may also realise that the purpose of this device is to ring at a specific frequency or tone. With both of these things considered, the name "The Bell" isn't only fitting, it is perfect.

Now it may seem likely that I specifically engineered this configuration to make this device appear as bell like as possible, but if we look at the experiments of Thomas Townsend Brown, we can see that he also adopted this exact configuration in his later designs.

Townsend Brown, Bell Configurations of Electrokinetic Apparatus

ELECTROKINETIC GRAVITATORS
TT BROWN

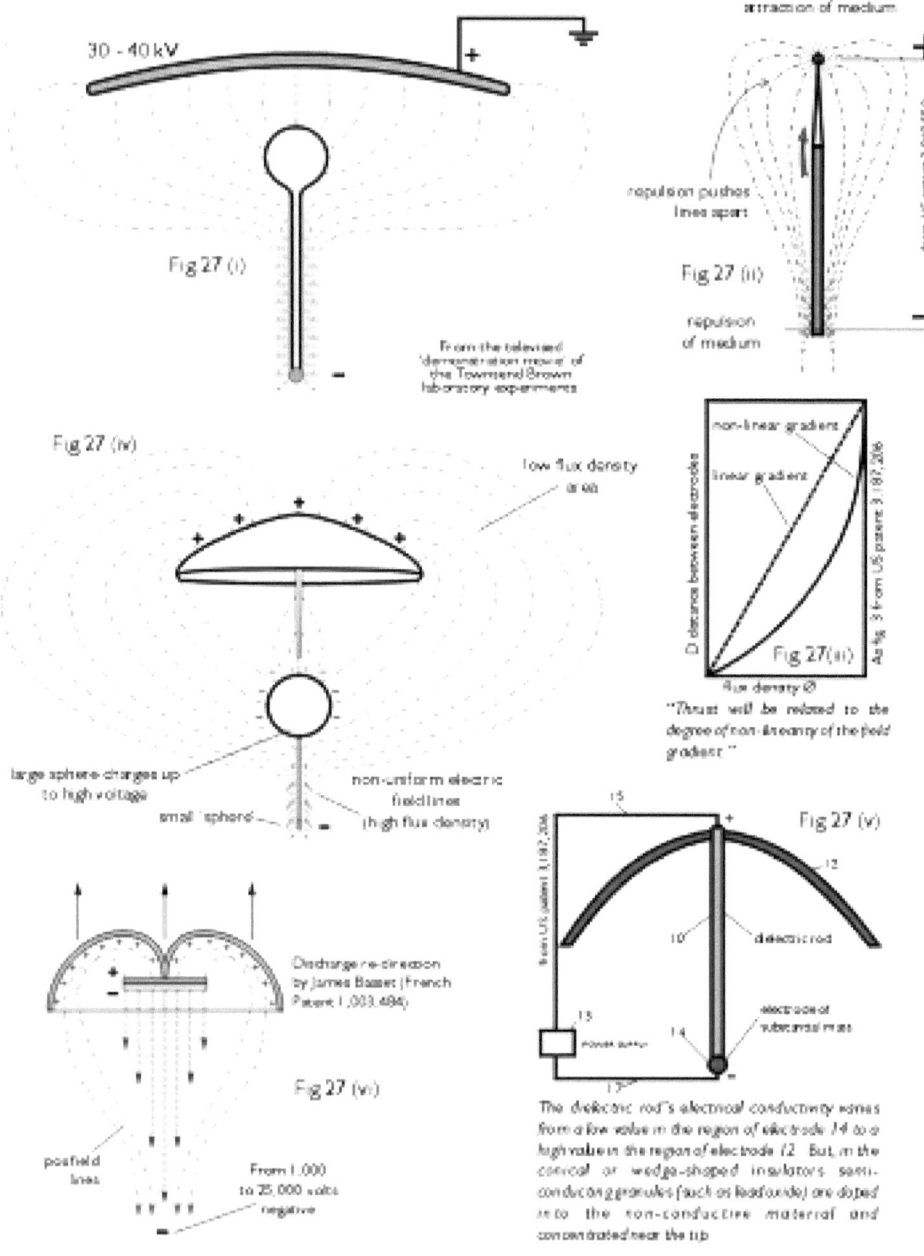

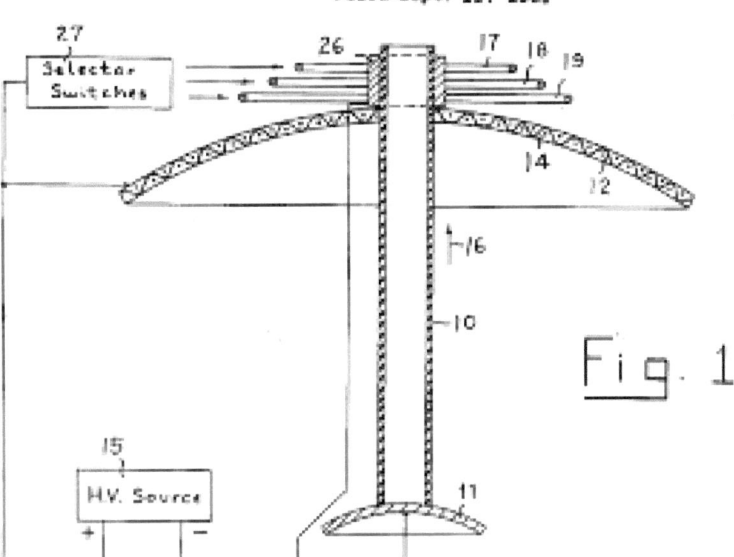

Fig. 1

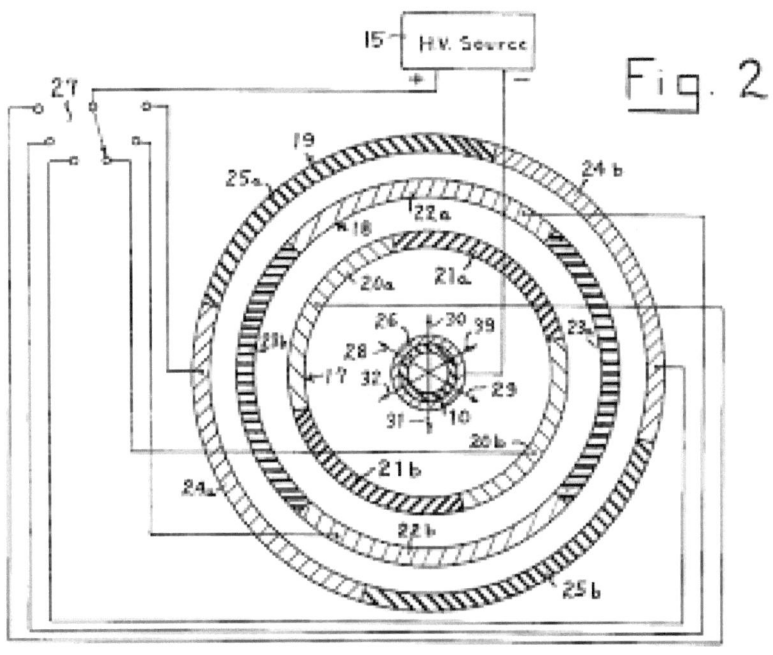

Fig. 2

Of course this all could be a coincidence, The Bell may never have existed at all or it may have existed but served a very different purpose. I have not explained the specifics of the stories surrounding this device just because the waters are so muddied with speculation, superstition and halfbaked theories that it would be a fruitless and exhausting endeavour to delve too deeply into it.

The coincidence seemed compelling enough to touch upon it in this book.

CHAPTER 3:

Bob Lazar

In the same year that I learned about the work of Thomas Townsend Brown, I also discovered the story of Bob Lazar and his accounts of working on Extraterrestrial gravity amplification systems.

Bob Lazar is a man that claims to have worked at a secret government weapons research site that was connected to the world famous "Area 51" at Groom Lake, Nevada.

In 1989, Bob made world news when he was interviewed by George Knapp on a local Las Vegas TV channel KLAS.
During his interview, Bob claimed to have worked on a genuine extraterrestrial spacecraft with a view to reverse engineering it.

He also claims that he saw 9 distinct different extraterrestrial spacecraft.

Unlike most eyewitness testimonies, Bob was able to provide detailed technical information about one of the vehicles that he saw.

Rather than retell Bob's story, I would encourage you to seek the information for yourself. There is a hell of a lot of it out there and he seems to be enjoying a return to the limelight lately so it should be very easy to find. A good place to start would be his own website at

https://www.boblazar.com

The reason why I am including him in this book is to highlight the differences between his own explanations of the technologies that the extraterrestrials use to manipulate gravitational fields and the similarities to my own explanations.

I will also say at the outset that I fully believe Mr Lazar's account of his experiences and will be writing about the things that he saw as absolute truth. My intention is not to debate the authenticity of his account, but to mine it for information.

At first it seemed as if my initial conclusions concerning gravity amplification directly contradicted Bob's descriptions of similar devices.
Bob's described 3 barrel shaped gravity amplifiers that rely on a mysterious fuel that he calls element-115, based on its atom weight.

He claims that the specific isotope of element-115 that he worked on had an unusual warping effect on the volume of Space and Time that surrounds it.

The word isotope refers to the arrangement of the sub-atom particles within the atom of this element. Most elements can be found with different isotopes, but in particular heavy elements like Plutonium and Uranium can come in a wide variety of different varieties. This means that their sub-atomic structures are configured differently, but their atomic weight or number is similar but not the same. Different isotopes tend to be more or less stable depending on their specific isotopes, for example

Uranium-238 is the most common isotope and can be found in nature, but it cannot be used in fission reactors whereas Uranium-235 can also be found in nature but is fissionable and is commonly used in nuclear reactors.

An isotope of element-115 has been artificially created in laboratories on Earth and therefore it can be found on the Periodic Table, but unfortunately the specific isotope of E-115 was very unstable and only existed for half a second. It is called Moscovium based on the location of the lab where it was first created (Moscow) in 2003.

The isotope that Bob seems to have worked with is apparently very stable and has an unusual effect on gravity. He believes that the presence of this particular isotope of E-115 is necessary for gravity manipulation technology in the case of the vehicle that he worked on.

I however, believe that he is mistaken. I do not think that this affects the authenticity of his accounts of working on extraterrestrial technology, rather than the accuracy of his predecessor's work. In other words, I think that they made a mistake when they attributed this technology's capabilities to the presence of E-115, although I think that their mistake was fully understandable.

If we consider the G.A.L, then a craft that uses this technology would not necessarily possess anything that we could identify as an engine. The gravitational effect is generated by the outer skin of any craft that uses this technology. Unfortunately the craft that Bob Lazar worked on does possess components that could be identified as engines and although I do believe that these components do function as part of the propulsion system of the craft, they are not the entirety of it.

I am referring to the barrel like structures that Bob described, deep in the belly of the craft.

In the interests of providing context, I will give a quick outline of Bob's own description of the vehicle that he worked on.

In Bob's eyewitness account, he mentions that he witnessed nine distinct, extraterrestrial craft, although he only worked on one. He named that craft, the Sports Model. The Sports Model was a metallic, disc shaped craft that was 52 feet, nine inches in diameter and 16 feet tall.

Sports Model Spacecraft

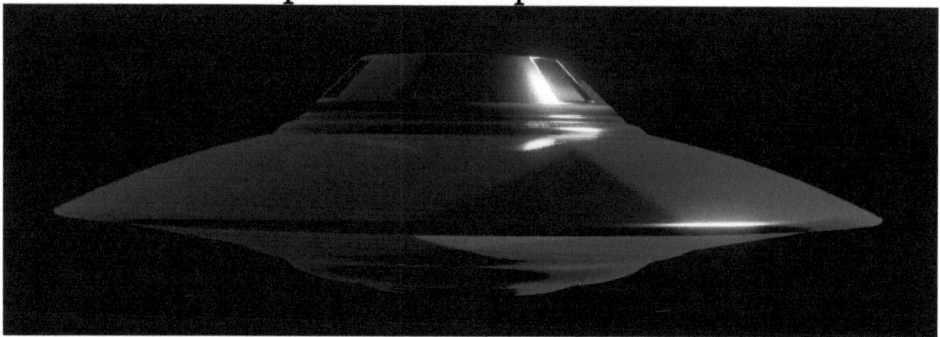

Internally, the vehicle was divided by three decks. Unfortunately Bob was never shown the top deck, so it's contents and purpose are unknown. The middle deck seemed to be the control centre for the craft, or bridge if you prefer. This deck contained three child sized seats, the tops of three gravity amplifiers and a single, basketball sized antimatter reactor that was in the middle of the floor, with a metal tube running between it and the ceiling. The lower deck contained three gravity amplifiers that were suspended from the ceiling by articulated gimbals.

Bob also noticed that there were no sharp edges to be found anywhere in or on the external surfaces of the craft. He tells us that everything appeared to have been milled from a single piece of metal and that everything looked as if it was slightly melted, like wax.

Sports Model Internal Devices

HOW TO BUILD A FLYING SAUCER

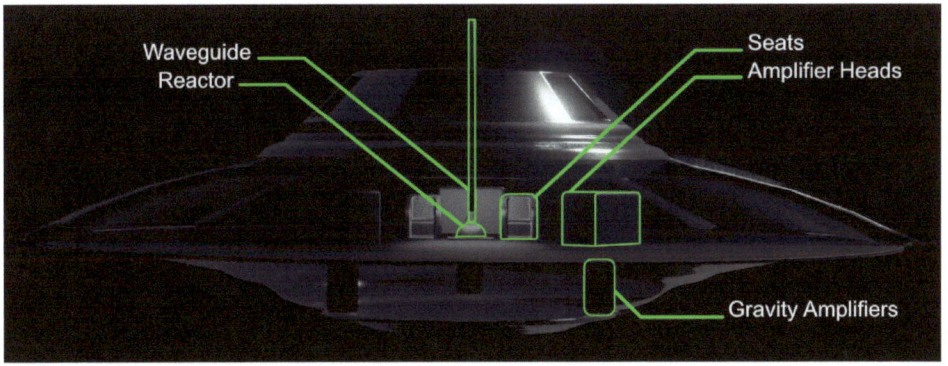

Bob believes that the gravitational effects of the craft are generated through the interactions of the small reactor on the middle deck and one of more of the amplifiers in the bottom section of the craft. In his description of how the craft operated, he tells us that Element-115 is transmuted into Element-116 by introducing an extra proton to its nucleus. This causes the material to Instantly decay, producing two anti-protons. This minute amount of antimatter is then funneled to a gaseous target and annihilated to produce power for the ship in the form of heat, which is then converted into electrical energy.

A side effect of this process is the production of gravity waves that are then led up through the long vertical tube, which Bob tells us is a gravity waveguide, and up through the roof of the craft. The waves are projected from the tip of the waveguide to then create a bubble of altered Space-Time that surrounds the craft, nullifying the effects of Earth's own gravitational pull.

Diagram Displaying Gravity Waves Counteracting Earth's Gravitational Field

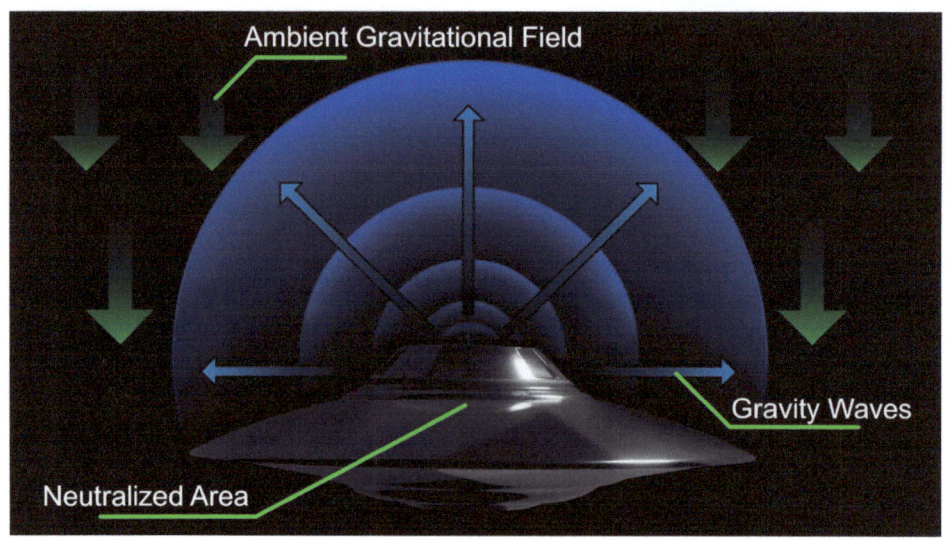

The gravity amplifiers are then free to move the craft laterally by generating a horizontal axis of gravitational force in the direction of travel.

Diagram Showing Gravity Amplifiers Utilized For Horizontal Propulsion

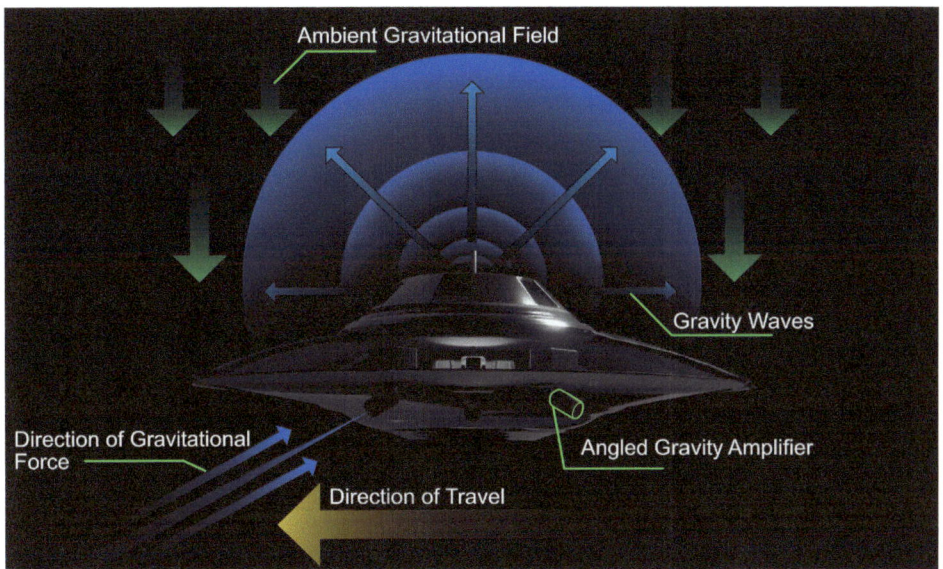

I personally don't believe that this hypothesis would work. The nullification of Earth's own gravitational pull, through the production of gravity waves seems unlikely. It would appear that the proposed method uses the directed flow of the gravity waves to counter the downward force of Earth's own gravitational field. To my mind the production of waves, barely affects the flow of the medium through which they are propagated. If we imagine a flowing river, were we to generate waves that move counter to the moving water, the opposite motion of the water would be minimal and very short ranged.

I think that it is much more efficient to focus gravity waves to a single point to generate a powerful but localised gravitational field.

If we look at the shape of the Sports Model, we can see that it has the appearance of a lense, perhaps with the intended purpose of focusing gravity waves at a single point somewhere below the craft.

Sports Model as G.A.L

Kinetic Force
G.A.L Unit
Gravitational Envelope
Focused Gravity Waves
Spatial Distortion

Another reason that I believe that the craft uses the same method of gravity manipulation that I outlined in the first chapter, is that when Bob describes observing this craft take off, he saw ionisation along the bottom surface of the craft. He also mentions that the craft would glow when flown at night.

When we combine this with the fact that Bob describes the surfaces of the craft as having no sharp edges, we can begin to see that when this vehicle is operating, it does so while carrying a huge electrical charge.

The presence of ionisation alone tells us that the skin of the vehicle contains a huge electrical charge when the vehicle is flown.

With Bob's description of the mechanisms that make this vehicle operate, we do not get an explanation for this excess charge and yet there must be a very good reason for its presence. Operating a vehicle like this would be potentially dangerous and wasteful. A rain shower would be enough to waste huge amounts of electrical energy and also make standing near to the craft potentially deadly. A single arcing spark from the outer surface of the vehicle could be enough to kill any bystanders.

We can assume that the reason that there are no sharp edges in or on the craft is to prevent this mishap from occurring, but under the right circumstances this excess of electrical charge could prove deadly.

I believe that the presence of this excess charge along with the shape of the craft are indicators that the skin of this craft functions as a Gravity Amplifier as outlined in the first chapter.

I do believe that the three smaller gravity amplifiers do function as Bob describes, but I think that he and his co-workers incorrectly attributed the vehicle's levitation to the craft's reactor.

I believe that the waveguide's actual function is as a kind of release valve, to vent off excess charge. This craft operates by containing and isolating a huge electrical charge, but it is also necessary to have a mechanism to lower that excess electrical potential. I believe that the waveguide is a conduit that allows excess charge to escape into the atmosphere by ionising the air above the vehicle. In space, this would be done by releasing an ionised gas from the tip of the pipe.

It is also worth mentioning that Bob also described a strange effect that emanates from the surface of the hemispherical reactor when it is operating. He noticed that when he tried to place his hand on the skin of the device, that there was a short range repelling force that would push his hand away. He described it as being similar to the repelling force of two matching magnetic poles that are held close to one another.

I believe that the hemisphere of the reactor is also a gravity amplifier. I think that this would be a useful method for isolating antimatter within the reactor so that it does not react with the reactor itself. When antimatter comes in contact with any matter, it instantly annihilates both itself and the matter that it comes in contact with. On Earth we use powerful magnetic fields to isolate it, but magnetic fields have very short ranges of effectiveness, whereas gravitational fields work at great distances. The ability to generate a gravitational field would allow much greater control of this dangerous substance.

If we imagine that the hemisphere of the reactor is a gravity amplifier, we can imagine the inner gravity waves focused upon a single point in space on the inside of the reactor, but if we look at the outer surface we would see a small gravity wave force moving outwards to.

This wave force would have very short range as the waves would quickly disperse as they are being generated by a convex surface, but close to the surface you would be able to perceive a repelling force as your hand pushes against the flow of the waves.

Reactor Gravity Wave Illustration

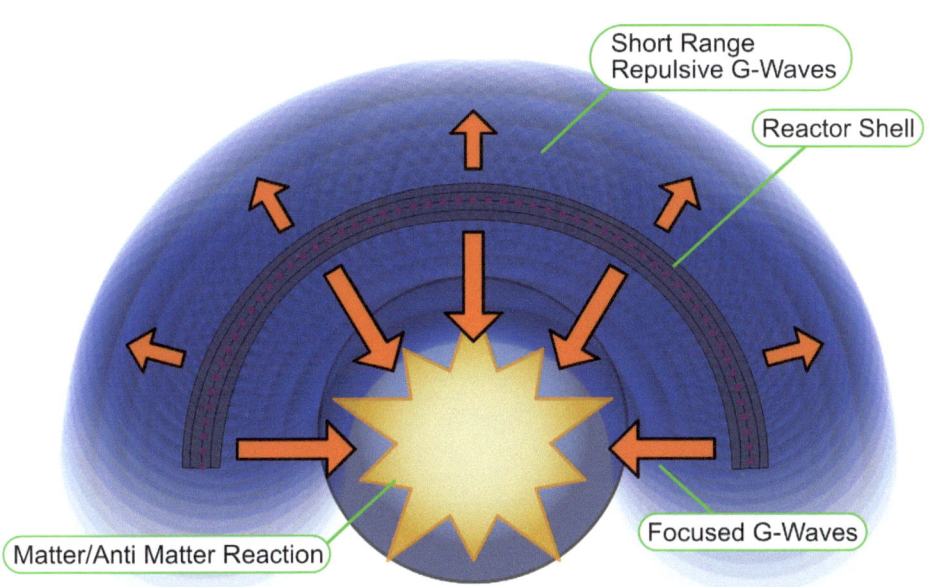

I have more thoughts concerning Bob Lazar's Sports Model, concerning the inner workings of the small gravity amplifiers, but maybe I will write those in a later book. For now I will move onto a different subject.

CHAPTER 4:

Close Encounter Evidence of Gravity Amplification

I will use this chapter to highlight UFO encounters that demonstrate evidence of G.A.L like gravity amplifier use.

The first and most obvious is Travis Walton.

Travis worked as a forestry worker during the seventies, but became world famous after he had a five day close encounter with a UFO that stated on November 5th, 1975.

On the night in question, Travis was riding home with his work colleagues in a truck when he and his friends all noticed a large, yellow, glowing object, hovering amongst the trees.

Travis decided to investigate the UFO and approached it on foot.

The object was disc shaped and clearly mechanical, it was twenty feet across and hovering twenty feet off the ground while emitting a high pitched sound.

As Travis got close to the craft his co-workers observed a beam

of bright blue-green light strike him in the chest and hurl him ten feet through the air. Travis describes the experience as if he was hit by a powerful electrical shock that almost immediately caused him to black out.
His co-workers observed this and convinced that Travis had been killed, fled in terror for their own lives.

Travis's story does not end there, as he recalls waking in an alien environment, encountering non-human beings and was returned to Earth five days later. His story is a fascinating one and can be found on his website at http://www.travis-walton.com/

Travis is a fantastic writer and I would recommend that you read his story, but for our purposes we will just focus on the first moments of his encounter.

Travis described the object that he encountered as appearing like two pie tins that were placed lip to lip with an upside down bowl sat on the upper pie tin. The object had a yellowish glow like hot metal and was making loud, low mechanical noises, like heavy machinery as well as a series of high pitched sounds.
Moments before the light struck him, he noticed that the noise suddenly got louder with a sound like large turbines powering up at the same time as the craft seemed to wobble on its axis.

Almost as soon as he registered this change, he was struck by a bolt of blue-green light that caused him to lose consciousness.

I believe from this description, that Travis was the unfortunate victim of the extraterrestrial equivalent to a road traffic accident. I would call it a hit and run incident, but rather than flee the scene, the occupants of the craft seemed to have nursed Travis back to health.

I think that this craft used the G.A.L method of gravity propulsion, but was experiencing some kind of technical problems. The presence of sounds indicates that the skin of the craft was not oscillating at the correct frequency. I believe that the occupants were trying to adjust the charge of the craft, so that the metal of the hull was vibrating at the harmonic resonance of the material that it was made from.

To do this they would have had to balance the charge to correctly tune the frequency of the gravity amplification surfaces. This would account for both the sounds and the glow of the object.

The cacophony of sounds would have been emitted by the hull of the craft as it searched for the correct frequency and the glow would have resulted from the presence of the huge electrical charge of the craft ionising the air around itself.

Diagram of Travis Walton's Encounter

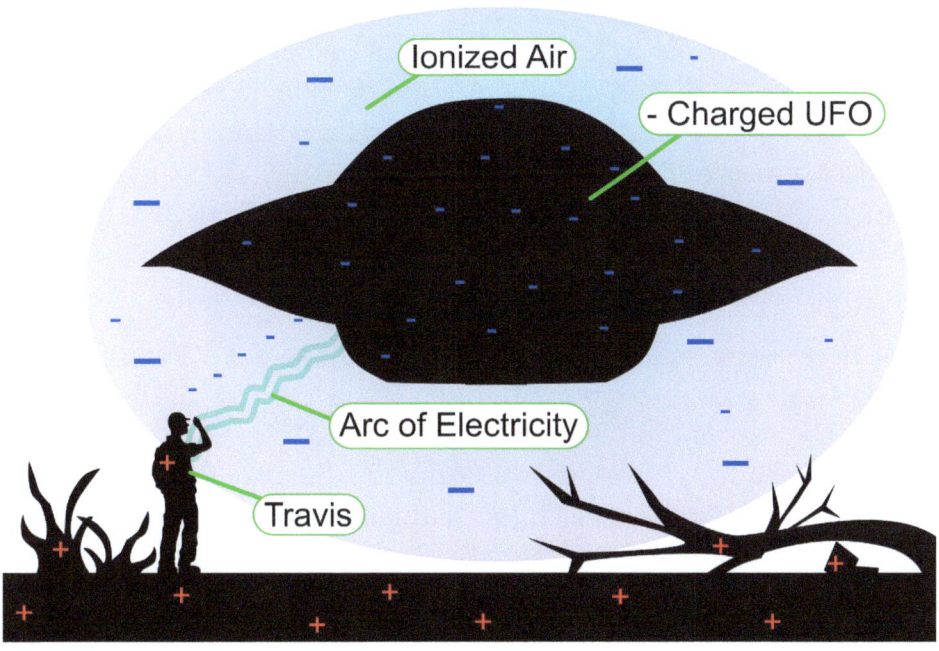

The Extraterrestrial's other option would have been too simply land in order to reset the whole system from the start. The problem with this is that a truck full of forestry workers had just pulled up and were not likely to go away if the ship landed.

Presented with two risky options, they tried their Luck and attempted to rebalance their drive system in the air. Unfortunately Travis got too close to their ship at precisely the wrong moment and as the highest, most conductive point in a cloud of ionised air, he received an arching bolt of electricity to his head and chest. Considering the distance that the bolt threw him, it seems unlikely that he would have survived his injuries, but luckily the occupants of the craft decided to help poor Travis and nursed him back to health over a span of five days. Travis can recall moments when he returned to consciousness aboard the extraterrestrial's craft, but that is a story that you can read in his book, which you can find at, http://www.travis-walton.com/book.html

I like to imagine a scenario similar to a driver that has unfortunately broken down in an illegal parking area. While this driver is trying to restart the engine of his car, he notices a traffic-warden/parking-attendant approaching, so he redoubles his efforts to restart his engine. He manages to restart his engine, but doesn't notice that his car is still in gear and so as the car starts, it also lurches forward, running over the poor traffic warden.

I feel that this scenario is very similar to the events that led to Travis's abduction. I would love to see a reboot of the movie "Fire in the Sky" which was loosely based on Travis's story. In the reboot, it could be filmed as a comedy of errors. Maybe that is too insensitive, Travis was injured and had his life turned upside down by the whole affair after all. It would be fun though, to see dialogue between the aliens as they discuss what the hell they should do after the accident. As I write this I'm imagining the alien characters from the TBS show "People of Earth," debating whether to leave poor Travis or help him. I miss that show and still haven't really accepted that it was cancelled after just two seasons.

That was quite a digression, but I rarely miss an opportunity to talk about "People of Earth."

If we look at common factors observed during UFO sightings we can find evidence that most of these objects use the G.A.L method of gravity propulsion.

The first correlation is that many of these vehicles are bowl, lense or disc shaped.
We know that this is the optimal shape for a craft that uses this system as it relies on the focusing of gravity waves below the craft.

The second correlation is simply that many of these craft also seem to possess a glowing aura. The presence of a colossal standing electrical charge is necessary for the operation of the G.A.L as it provides the kinetic energy that oscillates the atoms within itself, that then provides us with the gravity waves that drive the whole system.

Any object that contains a huge electrical charge will cause the air around itself to become ionised. This effect is generated by free electrons leaping from the surface of the object, the action of which provides us with a cloud of ions and a glowing aura.

The third correlation is that most of these craft seem to move silently. The G.A.L probably does make sound, but that sound is at such a high frequency that it would require special equipment to detect. The high frequency oscillation of the materials of the G.A.L system would most likely produce a deafening high frequency sound. This could also explain why some animals react badly to the presence of UFOs as many animals have the ability to detect much higher frequencies of sound. The presence of this high frequency sound isn't just undetectable by human ears, but it may have the effect of drowning out other ambient sounds around itself, projecting an eerie silent aura. Which could be why many witnesses feel compelled to comment on how silent these craft are.

The fourth correlation is that the G.A.L system would be the ideal instrument to draw complex patterns in fields of wheat. Whatever your feelings may be concerning the legitimacy of crop circles as a paranormal phenomenon, it should be noted that the G.A.L system of propulsion is singularly ideal for creating this unusual form of artist expression.

The G.A.L system relies on the generation of a localised, linear gravitational distortion that is located several meters under the central axis of the craft.

If this linear distortion was to be projected into the ground it would have the effect of pushing anything on the ground downwards. To produce a comfortable 1G of gravitational force for the occupants of the vehicle, we would expect that the G force felt at the focal point of the G.A.L to be many times stronger than this, creating a powerful directional, crushing force for anything unfortunate enough to find itself at this point in space.
The biological matter of any crop that finds itself in this cone of directional force would also find itself immersed in a bath of highly ionised air, which may also have the effect of softening it as it is slowly cooked by the presence of the highly energised ions.

I am not saying that all crop circles are really constructed by extraterrestrial vandals, but if just some are real, I am certain that this is the method that they used.

Gravity Amplifier used as Crop Stylus

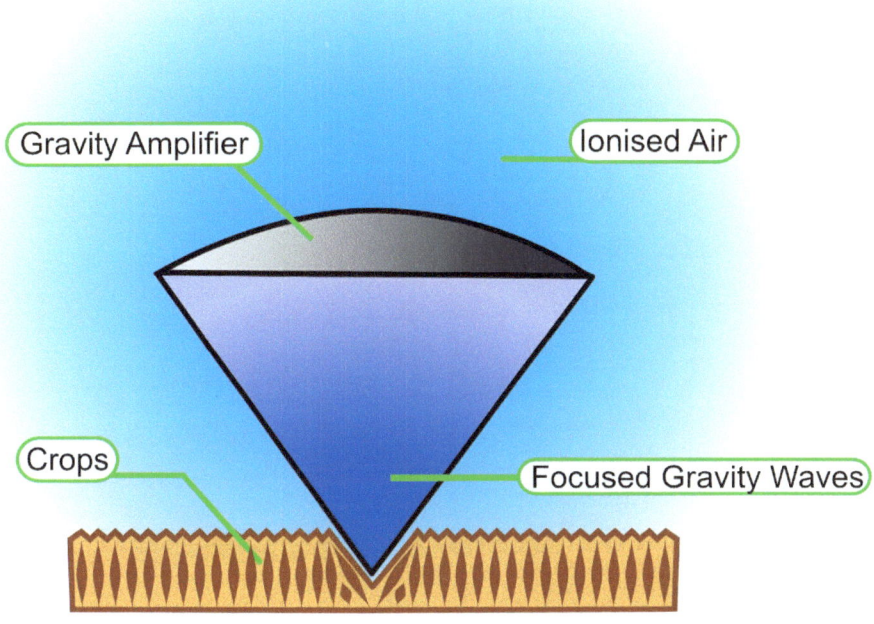

The fifth and final correlation is probably the most compelling for myself, but it is very specific and quite complicated.

Since the 1950s there have been several eyewitness accounts of disc shaped craft that seem to move vertically, in a falling leaf pattern of motion when either ascending or descending.

This has been documented many times and seems to be specific, to simple inverted bowl shaped craft.

This simple design would seem to indicate that these specific crafts are operating with a single Gravity, Amplifying Lense.

I believe that this eccentric manoeuvre may be a necessary side effect of operating a G.A.L system that relies on a single, amplifying surface.

Falling Leaf Motion of Single G.A.L Unit

HOW TO BUILD A FLYING SAUCER

A single G.A.L would only be able to generate vertical and horizontal thrust.

Horizontal thrust would be generated by simply applying an asymmetrical charge across the surface of the G.A.L. This has the effect of generating an asymmetrical distortion across the curvature of Space, which causes the axis of the G-force to tilt, providing horizontal motion and lift simultaneously.

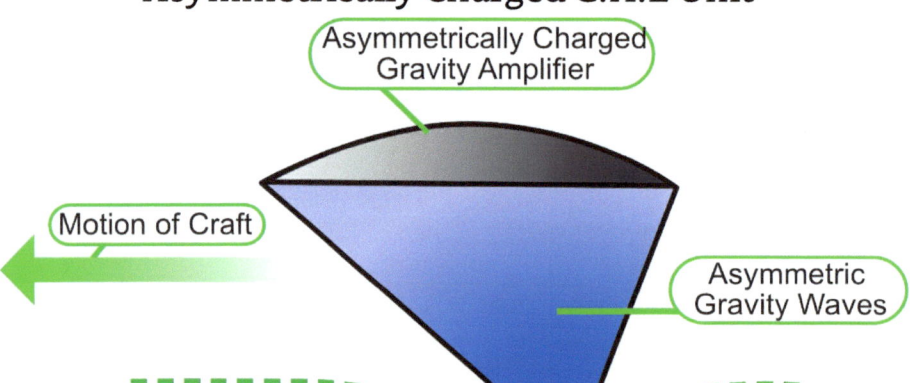

The G.A.L system would be extremely efficient for travel through the vacuum of space, but when operated within a planetary atmosphere, certain considerations must apply. The first consideration which we have touched upon several times is that the presence of an atmosphere is not an ideal electrical insulator. Whereas the vacuum of space is a perfect electrical insulator, which prevents the loss of charge from the surface of the craft, making the craft extremely efficient when travelling through the vacuum of space. When operating in atmosphere however, some loss of charge is experienced as the air around the craft becomes ionised, which also makes standing close to the vehicle extremely dangerous, as Travis Walton could testify.

Another side effect of operating in an atmosphere is that vertical travel may become difficult due to the extremely poor aerodynamic profile of the vehicle when traveling along its own vertical axis.

Another and less obvious side effect of atmospheric travel is that in order to travel vertically, the occupants of the craft must experience G-force greater than that of the planet that they are visiting. In order for the G.A.L system to render itself gravitationally buoyant, it must generate a force that is equal or greater than the gravitational field in which it finds itself immersed. This means that if a craft wishes to accelerate vertically at a high rate of speed, a strong gravitational field must be generated in order to punch through the atmosphere and counteract the force of the Earth's gravitational field. This also means that the pilots of these vehicles would have to experience powerful, crushing G forces if they wish to move quickly, perhaps to evade detection from witnesses.

An alternative method of rapid ascent may be to move rapidly along a more neutral axis, by traveling along a diagonal Z axis. That way craft would be imbued with a short period of rapid ascent.

The issue is that the craft would be unable to generate a straight diagonal vector of motion, due to the geometry of the gravity lense. No matter how asymmetrical the charge across the surface of the lense is, there will always be some vertical lift generated. This means that the diagonal ascent will always follow a curved path.

This isn't an issue when travelling horizontally, because the spatial distortion generated by the G.A.L unit is able to balance the horizontal thrust, with the lift necessary to remain at a constant altitude in order to move horizontally. This becomes a problem as soon as the craft tilts its axis as its own tilt is added to the tilt of the thrust of the G.A.L unit. The only way to counter this effect would be to charge the gravity lense symmetrically, which would cease all forward motion, and put us back to square one. This means that whenever the craft moves at a diagonal angel through a gravitational field, that it must always move along a curved vector.

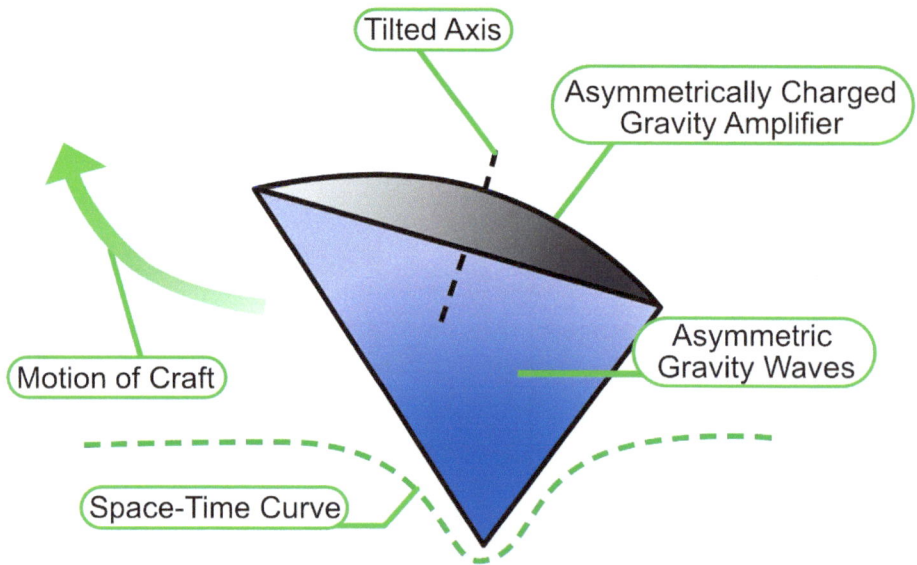

Diagram Displaying Curved Path of a Titled G.A.L When Ascending

This curved path would cause the angle of the craft increases until it eventually finds itself nearly at a vertical position. If the craft continues on this path, it would eventually follow a looped path and begin to descend. To prevent this from happening, When craft finds itself at a near vertical climb, the craft would simply rotate 180° along its own vertical axis and repeat the manoeuvre in the opposite direction. Another reason to avoid the craft from ascending vertically, even at a steep angle is that it would quickly become unbearable for the occupants inside the craft, because the G-Force inside the craft would have to be multiples of the downward force of the Earth's own gravitational pull.

To prevent the craft from falling at this point, the pilot would then quickly adjust the craft's angle back to horizontal and then perform the same manoeuvre in the opposite direction to gain more height.

Falling Leaf Manoeuvre Used to Ascend

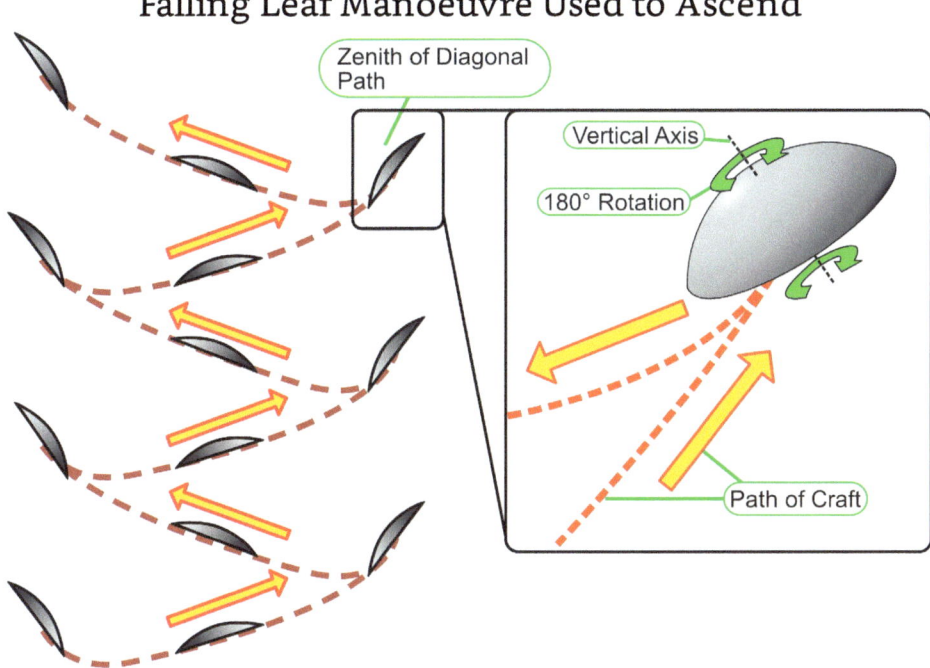

An almost identical manoeuvre could be used to descend, this may be necessary as a simple craft operating with a single G.A.L would not be able to descend faster than its terminal velocity. The terminal velocity is a term for the maximum speed that a free falling object can reach. This would be achieved by completely powering down the G.A.L and restarting it nearer to the ground, which may not be a quick process, making this method extremely hazardous.

A faster and much safer method of descent would be to tip the craft at a steep angle and use the crafts aerodynamic profile along with horizontal vector of gravitational thrust to descend along a steep angle. Once again we hit upon the problem that once we tilt the axis of the craft away from perfectly horizontal, we can only follow a curved path.

Diagram Displaying Curved Path of a
Titled G.A.L When Descending

HOW TO BUILD A FLYING SAUCER

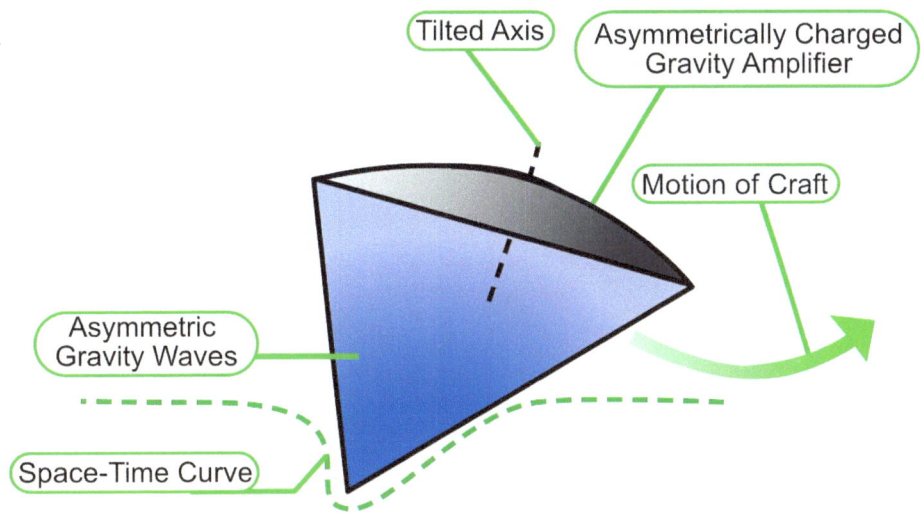

When we combine this factor with the fact that the air pressure under the leading edge of the craft also provides a lifting force as the craft descends, we can imagine that the vehicle would only be able to maintain a diagonal descend for a short period before its path drifts upwards, making the falling leaf manoeuvre essential yet again.

Falling leaf Manoeuvre used to Descend

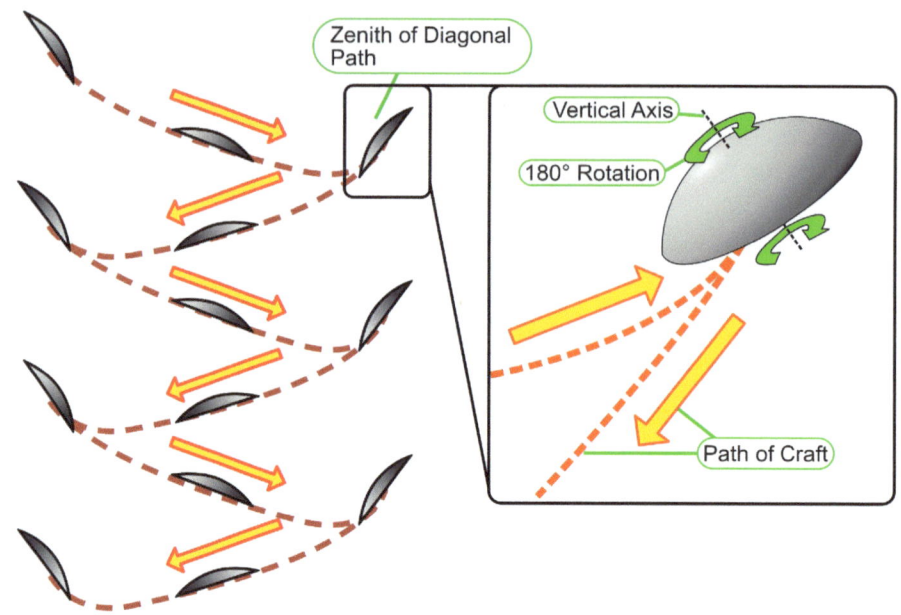

I mentioned earlier that to move vertically, a single G.A.L must generate more than 1 G of gravitational force. This is because the G.A.L system effectively shields everything between the G.A.L's point of focus and the outer, convex surface of the device from any Gravitational field that it immersed in by generating an extremely localized, linear gravitational field between these two points. This creates a cone of space within which any occupants are shielded from any Gravitational force outside of this conical envelope. This occurs even if the field generated is weaker than the ambient field surrounding it. This is because the dense cone of focused gravity waves within this envelope also has the effect of neutralising all external gravitational force.

You might think that this would prevent the occupants of these craft from feeling the effects of the G-Forces generated by the craft's motion and acceleration, but unfortunately in order to accelerate, the gravitational field that the craft produces, must match the G forces that would normally be produced by performing these manoeuvres using more conventional means. Which is why vertical acceleration requires that the force generated by the G.A.L system must be greater than the ambient gravitational field in which it is immersed. This means that performing high G manoeuvres subjects the occupants of these craft to the identical G forces that they would experience in conventional aircraft, because the craft's own gravitational force is always equal or greater than the G force that it is immersed. This includes G forces generated by rapid horizontal acceleration.

This seems to contradict many eyewitness accounts of the capabilities of extraterrestrial spacecraft, but I believe that this kind of constraint only applies to simple, single G.A.L systems of propulsion.
Luckily there is a solution to this problem and we will now look at G.A.L configurations with much greater capabilities than the single amplifying, lens system.

CHAPTER 5:

Multiple G.A.L Configurations

We have touched upon the limits of this technology though observation of simple iterations of the E.T's versions of it.
The falling leaf manoeuvre offers is a glimpse into necessary strategies to deal with these limitations, but not all UFOs seem to share these limitations. Many UFOs seem to be able to perform manoeuvres that seem to defy the laws of physics.

I believe that one engineering solution to achieve these incredible physics defying manoeuvres, is to use self contained gravity amplification units within the internal space of your craft, as Bob Lazar observed on board the "Sports Model."

Another solution which does not depend on accessibility to rare, heavy elements is to simply use several gravity, amplifying surfaces.

This is done by simply stacking them with differing focal points. I believe that the "Sports Model" also used this solution, perhaps as a failsafe in case there is a malfunction of the internal amplifiers. I believe that this is the case, based on the shape of the craft.

If we look at the Sports Model we can see from its side profile that it appears as if a small bowl has been stacked upon a larger dish. In fact if we were to remove the outer apparatus on this upper section of the craft's hull, I believe that we would find a smooth hemisphere beneath, lending the craft the appearance of a symbol. Anyone that is familiar with general shapes of UFOs will recognise that this symbol like shape is very common. I believe that this is because this configuration is extremely useful for craft that are required to navigate through planetary atmospheres, while also imbuing them with the ability to perform manoeuvres that would generally be considered impossible at our current level of technological advancement.

Speculative Image of Sports Model

If we remember that anything that is placed within the gravitational envelope of a G.A.L is shielded from external gravitational forces we can see that anything placed within the envelope of the upper G.A.L would be shielded from the forces generated by the lower G.A.L.

To prevent G-wave interference, the lower G.A.L is a flattened ring shape rather than the more efficient dish shape. This allows the focused waves from each G.A.L to move through a volume of space without interference from waves generated by the other G.A.L.

This means that if we place the bridge or control room of the craft within the envelope of the upper G.A.L, then anything within this volume of space would be protected from drive G-forces and inertia generated by the lower G.A.L.

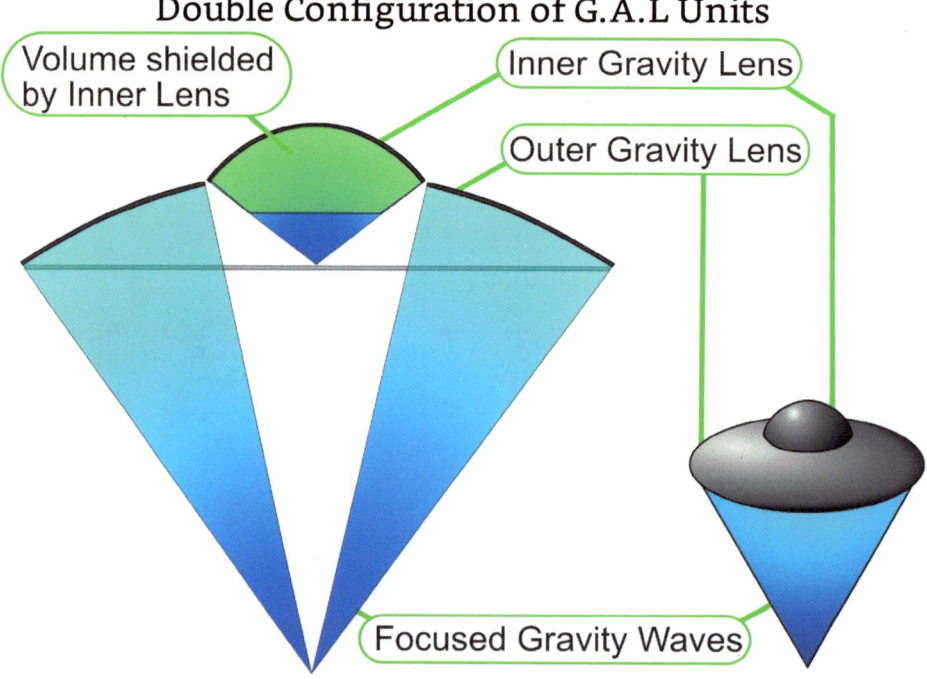

With this configuration of lenses, we are able to build a vehicle capable of generating G-forces that would usually be lethal for all known biological life, while also providing a bubble of protection from those forces.

To perform horizontal manoeuvres, the lower lens could be constructed from several plates, radiating outward from the hub of the craft separated by a non-conductive material. This would allow the lower G.A.L unit to generate an asymmetrical gravitational distortion, by increasing the amplitude of the oscillations of individual plates within the disk.

Double Stacked G.A.L Unit with Radial Plates

If we briefly return to Travis Walton's description of the craft that he encountered in 1975, he mentions that on close inspection, that the glow of the outer ring of the craft is divided into rectangular sections by darker seams that run from the edge of the disk toward the hub.

From this description we can see that the solution for horizontal manoeuvres outlined above, may have been adopted by the pilots of the craft that Travis encountered.

The problem here is that the surface that Travis was observing was on the underside of the craft, which would create conflicting

G-Waves with the upper disk when powered up. That is unless it was only ever used in tandem with the top, inner G.A.L unit to allow the craft to descend vertically in a controlled manner.

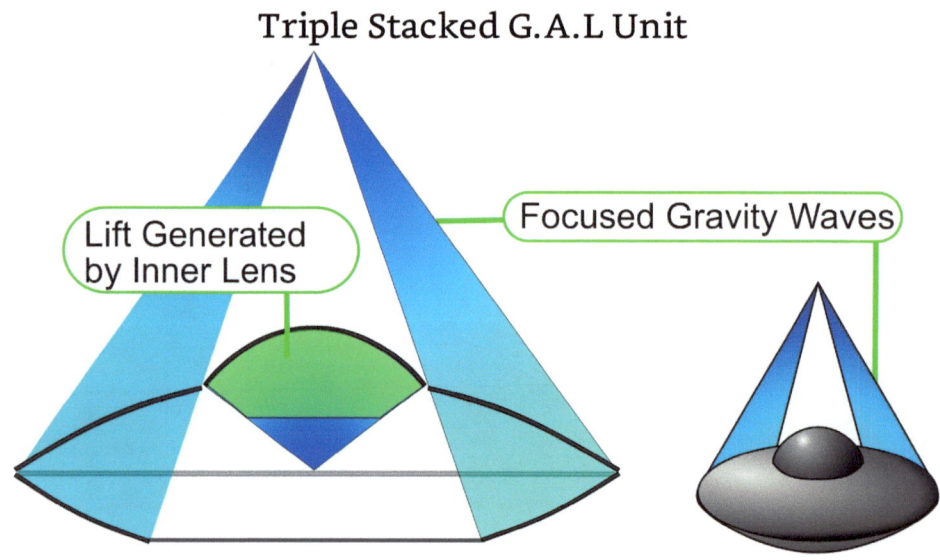

This configuration could also be used to also generate smooth, horizontal manoeuvres, although the occupants would be subject to some lateral G-forces.

Triple Stacked G.A.L Unit Utilised for Horizontal Travel

HOW TO BUILD A FLYING SAUCER

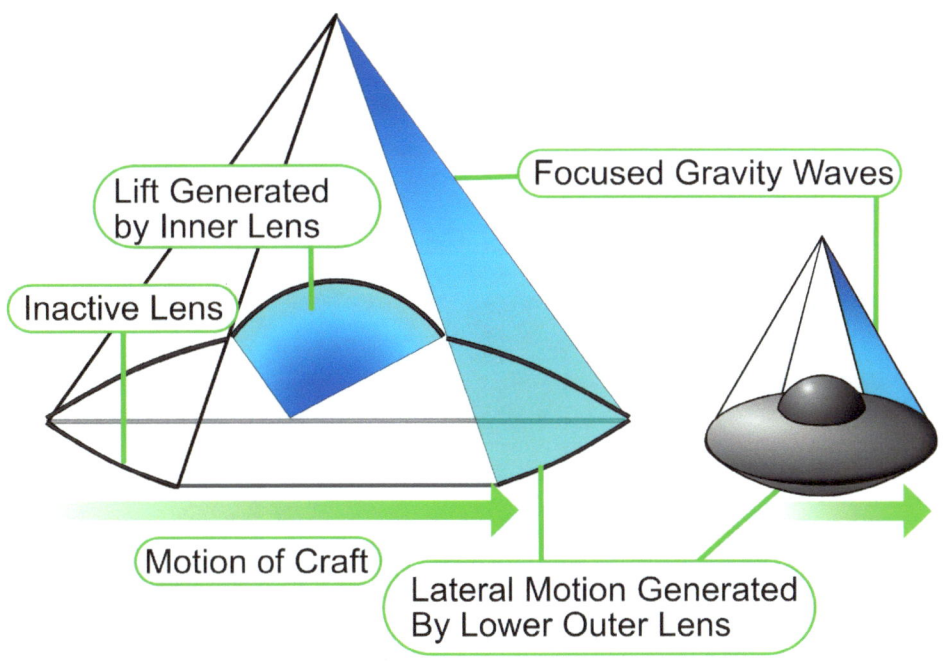

If we also extend the hull below the craft, with the intention of attaching a module containing small and powerful focused gravity amplifiers, we would then have the freedom to perform right angled turns at velocities that would normally reduce the unfortunate pilots into a jelly, with minimal risk to those pilots.

Triple Stacked G.A.L With Ventral Mounted Gravity Amplifier Bay

At this stage if we then also mount equipment to the outside of the upper lens, perhaps containing air intakes to draw in gases to be used for fuel, life support systems and sensor equipment, we would have something that looks very much like the craft Bob Lazar worked on during his time at Groom Lake.

Triple Stacked G.A.L with Instrument Hood Mounted to Upper G.A.L

It would make a lot of sense to mount delicate equipment to the upper section of the craft, because it would be within the gravitational envelope of the 1st G.A.L unit which is intended to provide a safe environment for the occupants of the craft. Close to the outside of the lens we would observe a short range repelling force, but the concentration of the G-Waves at this location would shield our equipment from external G-forces.

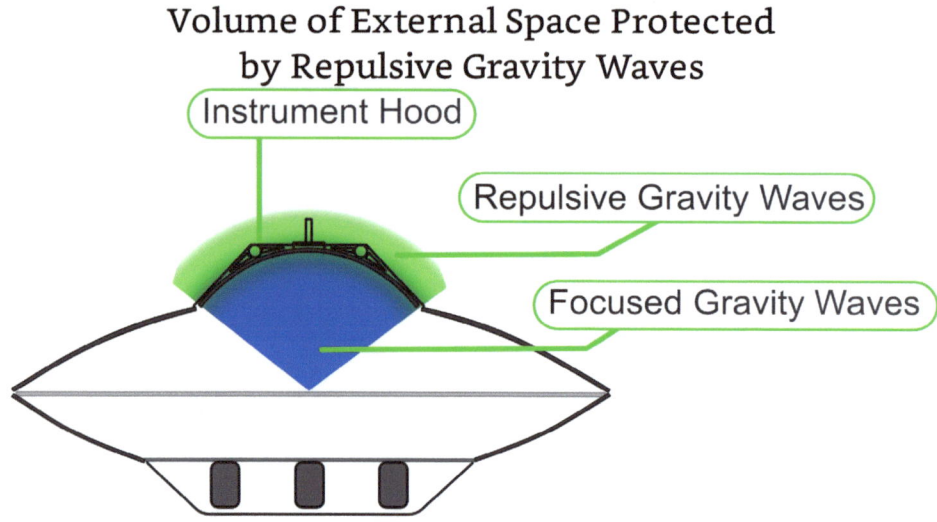

At this stage we might realise that the presence of the focused Gravity Amplifiers mounted in the ventral (underside) extension of the craft, render the lower G.A.L unit completely redundant. So we would probably end up with something that looks like this.

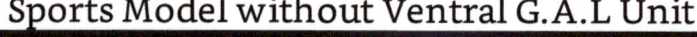

I built the CG model in the image above using the original diagrams created by Bob Lazar, which could indicate that we are on the right track concerning the role of the shape of the outer hull of these types of craft.

So at this stage of the book, we have explained why many extraterrestrial vehicles seem to follow a specific design constraint and we have also covered a wide range of the saucer like configuration designs, from the simplest to more complex multi role vehicles, like Bob Lazar's Sports Model.

We have been able to explain very specific manoeuvres and phenomenon associated with these vehicles, from their tenancy to glow, to their ability to perform seemingly impossible manoeuvres.

At this stage, I will talk briefly about a very recent development involving a piece of debris that seems to have come from an actual extraterrestrial vehicle and explain why this small scrap of metal has inspired me to release this information to the public.

CHAPTER 6:

Tom Delonge's Metamaterial

Recently, as in the date that I am writing this, October 2019, the news regarding UFOs has been dominated by speculation regarding a small fragment of metal that is believed to be linked to a very significant event in history.

The event that I am referring to is the Roswell crash and recovery.

Most already know the story, but a brief outline is that in the summer of 1947, an extraterrestrial disc shaped craft is believed to have crashed into a cattle ranch, close to the small American Town of Roswell in the state of New Mexico. At the time the story made local headlines. Rather than retell the whole story I will transcribe the news story from the "Roswell Daily Record" for the day of July 8th, 1947.

RAAF CAPTURES FLYING SAUCER ON RANCH IN ROSWELL REGION

No Details of Flying Disk Are Revealed

Roswell Hardware Man and Wife Report Disk Seen

The intelligence office of the 509th Bombardment group at Roswell Army Field announced at noon today, that the field has come into possession of a flying saucer. According to information released by the department, over authority of Maj. J. A. Marcel, intelligence officer, the disk was recovered on a ranch in the Roswell vicinity, after an unidentified rancher had notified Sheriff Geo. Wilcox, here, that he had found the instrument on his premises. Major Marcel and a detail from his department went to the ranch and recovered the disk, it was stated. After the intelligence officer here had inspected the instrument it was flown to higher headquarters. The intelligence office stated that no details of the saucer's construction or its appearance had been revealed.

Mr. and Mrs. Dan Wilmot apparently were the only persons in Roswell who seen what they thought was a flying disk. They were sitting on their porch at 105 South Penn. last Wednesday night at about ten o'clock when a large glowing object zoomed out of the sky from the southeast, going in a northwesterly direction at a high rate of speed. Wilmot called Mrs. Wilmot's attention to it and both ran down into the yard to watch. It was in sight less then a minute, perhaps 40 or 50 seconds, Wilmot estimated. Wilmot said that it appeared to him to be about 1,500 feet high and going fast. He estimated between 400 and 500 miles per hour. In appearance it looked oval in shape like two inverted saucers, faced mouth to mouth, or like two old type washbowls placed, together in the same fashion. The entire body glowed as though light were showing through from inside, though not like it would inside, though not like it would be if a light were merely underneath. From where he stood Wilmot said that the object looked to be about 5 feet in size, and making allowance for the distance it was from town he figured that it must have been 15 to 20 feet in diameter, though this was just a guess. Wilmot said that he heard no sound but that Mrs. Wilmot said she heard a swishing sound for a very short time. The object came into view from the southeast and disappeared over the treetops in the general vicinity of six mile hill. Wilmot, who is one of the most respected and reliable citizens in town, kept the story to himself hoping that someone else would come out and tell about having seen one, but finally today decided that he would go ahead and tell about it.

The announcement that the RAAF was in possession of one came only a few minutes after he decided to release the details of what he had seen.

The very next day another article was released that completely dismissed the possibility that the object that was recovered was a flying saucer as had been stated the previous day. The explanation given, was that the object was merely a weather balloon.
It is generally believed, in Ufological circles at least, that the ori-

ginal article reflects the truth and the second one was an attempt by the US military, intelligence department, to put a lid on the whole story.

In the years since the incident, there have been over 200 individual testimonies recovered from eyewitnesses that all claim to have observed direct evidence that the object was not of terrestrial origin.

One possible witness may have secretly obtained several pieces of the wrecked craft shortly after the crash, one of which may now be in the possession of Tom Delonge.
I am referring to the small piece of metal that has been getting a lot of public attention lately. This fragment is commonly referred to as the "Metamaterial."

This single piece of metal seems to have had a long journey, from the crash site in Roswell, New Mexico, to the science labs of "To The Stars Academy ".
If you consider the possibility that this thing is really a piece of an extraterrestrial spacecraft, that journey could potentially be considerably longer.
If we focus on the terrestrial leg of that journey, then we may learn something concerning its origins.
This small piece of metal came into the possession of Emmy award winning journalist, Linda Moulton Howe, in April of 1996. It was sent along with several other pieces with a letter that explained the original origin of these fragments of metal.
The letter explains that these pieces of metal were pulled from the wrecked hull of an extraterrestrial spacecraft, that was retrieved from the desert in Roswell, New Mexico, in the summer of 1947.

The writer claimed to be a soldier, that wished to unburden himself of a long family secret. Apparently his grandfather was one of the soldiers that were sent to secure the original wreckage at Roswell in 1947.

While guarding the wreckage at Roswell Air Base, before it was sent to the more secure location of Wright Patterson Air Force Base, he was able to snap off several charred pieces from a damaged section of the craft and smuggle them home after his shift ended.

Years later, he told his grandson this fantastic story before he passed away in 1974.

In the mid nineties, the subject of UFOs was getting a lot of public attention, thanks to TV shows like "The X-files." This rise in popularity of the subject led to the rise of radio shows like "Dreamland," with Art Bell and regular guest, Linda Moulton Howe.

Our nameless soldier, saw this as an opportunity to unburden himself and sent fragments of the metal that his grandfather had stored away since seizing them 49 years earlier to the radio studio "Coast 2 Coast AM," for the attention of Art Bell and Linda Moulton Howe.

Since receiving these fragments, Linda has worked hard to unlock their secrets by exposing them to a battery of scientific tests, at various laboratories.

She was able to discover that one of the larger pieces, roughly one inch in size, was constructed from. 26+ layers of alternating Magnesium/Zinc alloy and Bismuth.

The layers are incredibly thin, the thickest being only 200 microns thick at its widest point. She discovered that the Magnesium/Zinc layers were between 100 and 200 microns thick, while the Bismuth layers were only around 1 to 4 microns thick.

All of the pieces had a rippled pattern, seeming to indicate that they had been exposed to extreme heat or impact, during the crash.

In July of 2019, "To The Stars Academy" acquired several of these fragments, with a view to unlocking their secrets.

The largest fragment, is currently being tested again, which seems to have drawn the interest of the world's media.

I believe that they may be about to make some Earth shattering discoveries related to these materials.

In a recent interview on the "Joe Rogan Experience" podcast, Tom Delonge the founder of the "To The Stars Academy" expressed an interest in exposing these fragments to high frequency electrical currents.
I believe that this method of experimentation will not have much success, but it may be enough to put them on the right path.

I believe that the material was designed to contain charge, rather than conduct it, but they may be able to observe some interesting effects if they expose it to a particularly high voltage, although this may lead to further damage of the internal structure of the materials.
Neither Magnesium or Zinc are particularly good conductors, and Bismuth literally becomes a resistor when exposed to magnetic fields, but the low conductivity of the Magnesium-Zinc alloy may be enough to agitate the molecules within the material, enough to observe some strange effects. I believe that the low conductive alloy was specifically designed to oscillate at high frequency when charged. It is like a finely tuned musical instrument, and even incorrect exposure to current may be enough to lead to some interesting observations.

I think that it is entirely possible that this humble chunk of metal, could lead to a discovery that could completely transform us as a species.

When we apply my own hypothesis concerning gravity manipulation, the structure of this material seems ideally suited.

The metal alloy, although terrible as a conductor, would be a very good choice for containing a large static charge. I suspect that the Magnesium-Zinc alloy may be particularly conducive for molecular oscillations. The thin layer of Bismuth, would be an ideal dielectric between the alternating charged Magnesium-Zinc layers.

The rather vague number of 26+ layers, may be due to damage that the metal received during the crash. So we can probably assume that the original number was 27 layers with one layer that has mostly being broken away or scraped off during the crash. We could imagine these 27 layers as 7 stacked capacitors.

Each of the alloy layers would contain alternating positive and negative charges. The thinness of these layers ensures that the atom's oscillations, only occur on a vertical or Z axis. The Bismuth layer creates an insulating barrier between the alternating charges of the Zinc/Magnesium alloy layers and the extreme thinness, allows for maximum magnetic permeability between the alternatingly charged layers. This is important to creating the maximum amount of motion along the Z axis, for the production of gravity waves.

Metamaterial Diagram

HOW TO BUILD A FLYING SAUCER

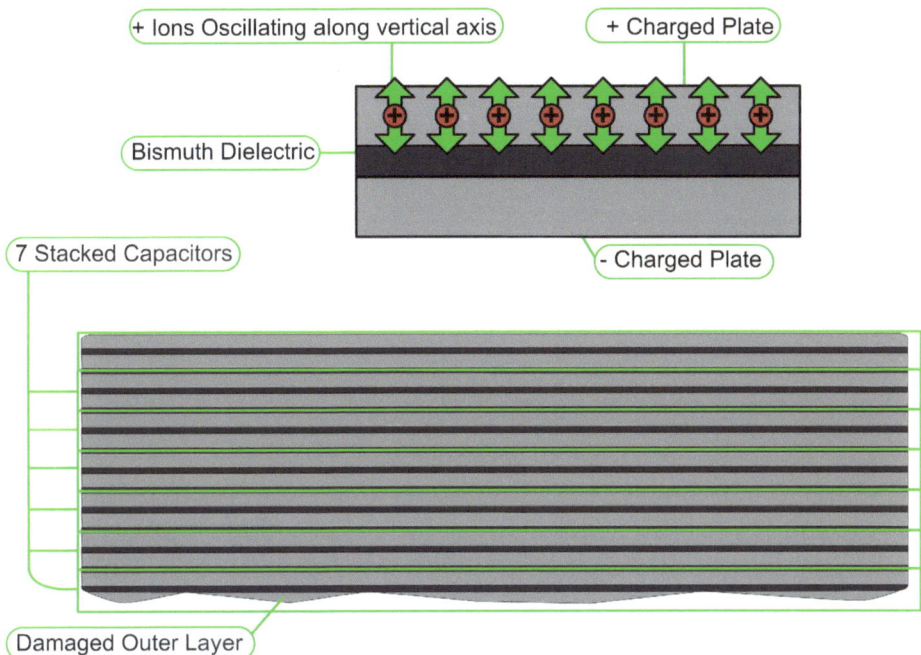

I am surprised that no-one has considered that this material was designed to contain large alternating electrical charges yet, or maybe the damage to it has destroyed its ability to contain alternating charges between layers. Any leakage between the layers would create tiny areas of current flow that would create extreme hot spots within the structure of the metal and effectively cook it.

I think that the best way forward, would be to try to recreate this material to observe it while powered up with alternating charges.

I believe that if they are able to accurately recreate the extremely thin layers of these materials and maintain a closed barrier between the layers, that when powered with a sufficient charge, that they will observe an area around it that is unaffected by the Earth's gravitational field. If they also shaped their newly constructed gravity Amplifier into a dish shape, they would be able to create and direct, linear gravitational fields.

This discovery could create a new phase of human expansion out into space.

I sincerely believe that we are on the verge of witnessing the most significant event in all of recorded human history. The expansion of our species out into space and contact with a variety of extra-terrestrial species. Part of the reason that I believe this is because up until right now, I have been unable to write this book.

What I mean by that is that I have tried many times, but something like writer's block prevents me from progressing shortly after I start to write. I would say that it is similar to writer's block, but it is much much stronger. The thought of continuing to write becomes unthinkable and I am always prevented from continuing from writing by this powerful compulsion.

I started to write this book about a month ago, and I am very close to completing it. It took more than twenty years to start, but now I find myself pouring the words into the page as if I am merely transcribing them rather than writing an original work.

I will take this as a sign that the time is right and will simply continue until the book is ready and quietly release it myself.

I have no idea what effect this information will have on the world, but I am sure that now is the right time to release it.

I will end here by sharing a variety of different configurations of the G.A.L system, using the information within this book to help uncover why certain types of UFO appear to be in the variety of shapes that we commonly observe.

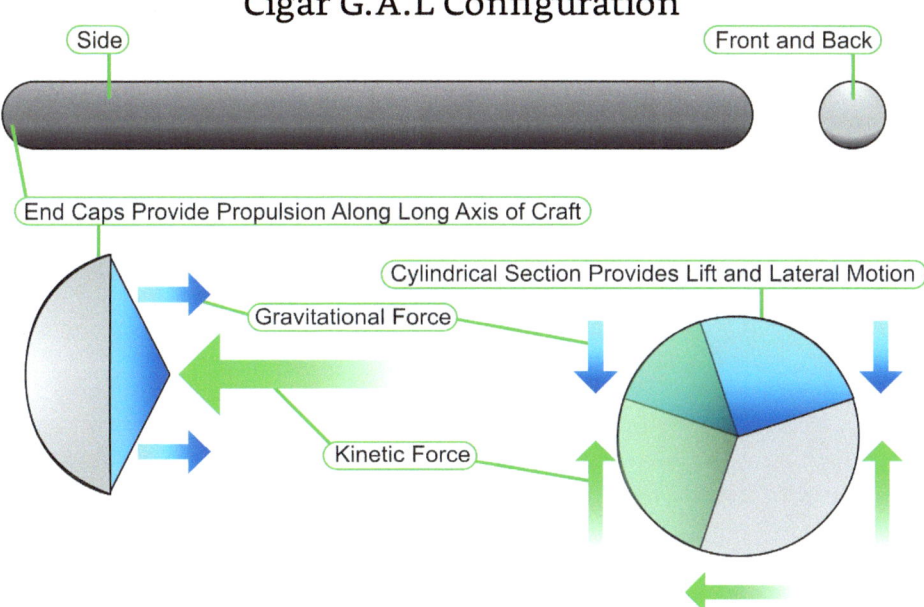

Tic-Tac or Lozenge G.A.L Configuration

(Side) (Front and Back)

Almost Identical to Cigar Craft with reduced Internal Capacity

End Caps Provide Propulsion Along Long Axis of Craft

Cylindrical Section Provides Lift and Lateral Motion

Gravitational Force

Kinetic Force

Spherical G.A.L Configuration

HOW TO BUILD A FLYING SAUCER

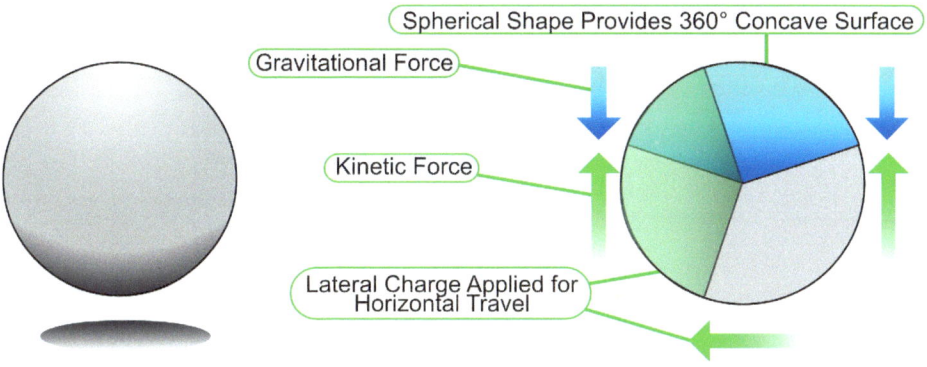

Egg Shaped G.A.L Configuration

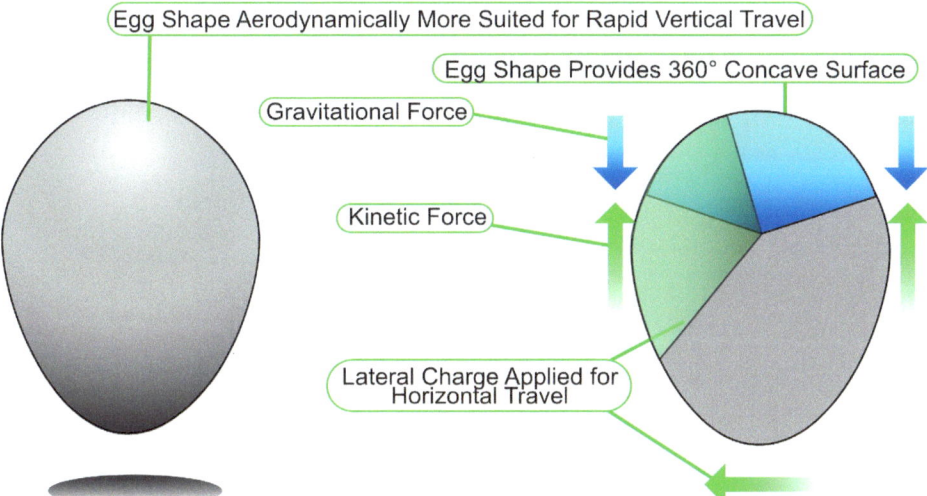

Triangular Configuration of G.A.L Units

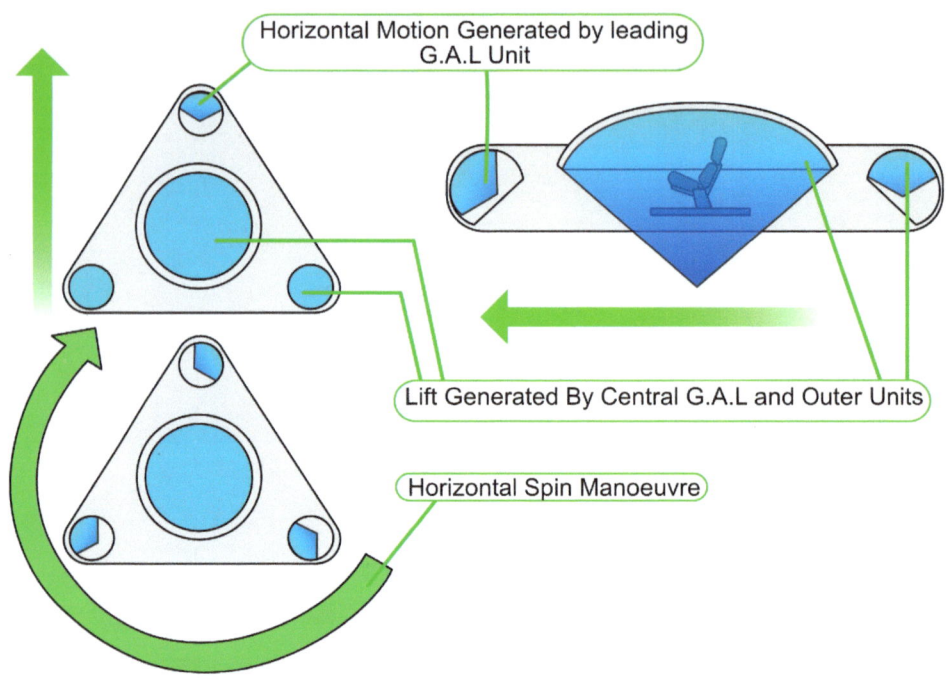

Rectangular Configuration

HOW TO BUILD A FLYING SAUCER

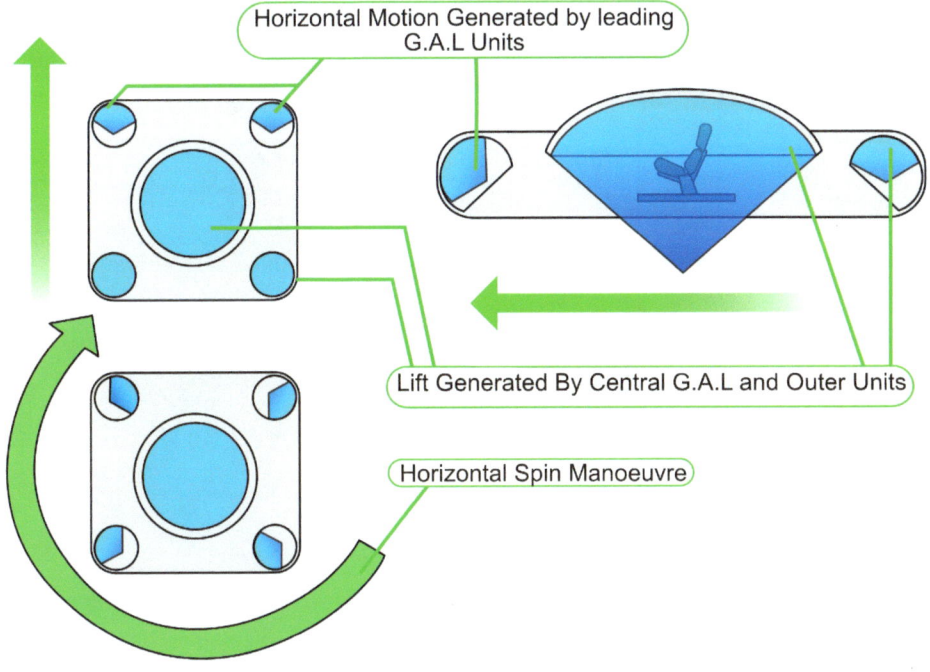

Dumb Bell Configuration

MICHEAL ALANS

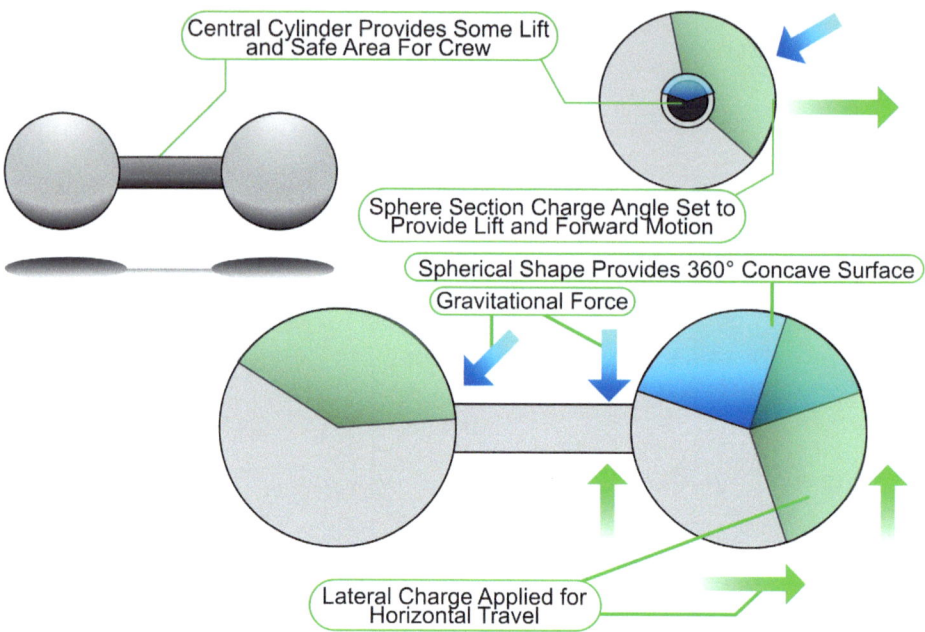

AFTERWORD

I hope that you enjoyed my book, as I stated at the start, you can read about my experiences with extraterrestrials in my first book, "Alien Revelations."
It is not necessary to understand the contents of this book, but in case you are curious, you can find it here, <u>Alien Revelations</u>
If you enjoyed my amateurish attempts to explain complex scientific information that I received from my time with extraterrestrials, then you should enjoy it. You will find my explanations for the birth of our Universe, the existence of the forces of Gravity and Electromagnetic radiation, the nature of consciousness and you will even find out what the E.Ts told me happens to us when we die.

The main reason for adding this afterword was to say that there is one part of this book that troubles me. At the start I stated I believe that it is possible to generate gravity waves by oscillating matter so fast that it is moving at a rate that is a measurable fraction of the speed of light.
I do believe this, but I don't know it for sure.
I am sure that the rapid oscillation of the matter does create gravity waves, but I am not certain that the effect is generated by the motion of the matter in relation to the speed of light. I came to that conclusion by myself.

I am usually able to judge how likely information is to be true, based on a strong feeling of déjà-vu. The conclusion that the matter within the materials that are used to construct gravity amplifiers are oscillating fast enough so that their mass is affected by

their relative motion based on Einstein's equation does not illicit the feeling of déjà-vu. This does not necessarily mean that it is inaccurate, but possibly my lack of knowledge of Einstein at the time that I received this information may have prevented the extraterrestrials from using this information.
The extraterrestrials are extremely adept at sharing general and sometimes abstract information, but anything that requires language or concepts that you simply have not learned yet is unavailable to them.

They use our minds as a kind of lexicon from which they can draw information, but anything that is absent is simply unavailable for them to use.
They also have difficulty expressing numbers, so I do know that the atoms within the plates of the G.A.L, must oscillate really, really fast, but the specifics of that rate are unclear.
When I combined this knowledge with the fact that Space and Time are affected by sufficiently rapidly moving masses, it seemed obvious to me at least that this was the source of the gravity waves. Unfortunately I have very little confidence in conclusions that I have drawn by myself and so I have added this little disclaimer at the end of my book.

I hope that you enjoyed my book despite my lack of confidence in my own conclusions.

www.ingramcontent.com/pod-product-compliance
Lightning Source LLC
Chambersburg PA
CBHW040318220526
45473CB00009B/2479